ELEMENTARY BLUEPRINT READING

FOR MACHINISTS

5th Edition

ELEMENTARY BLUEPRINT READING
FOR MACHINISTS

5th Edition

DAVID L. TAYLOR

DELMAR

THOMSON LEARNING™

Australia • Canada • Mexico • Singapore • Spain • United Kingdom • United States

DELMAR

THOMSON LEARNING

Elementary Blueprint Reading for Machinists

David L. Taylor

Vice President, Technology
and Trades SBU: Alar Elken

Editorial Director:
Sandy Clark

Senior Acquisitions Editor:
James Devoe

Senior Development Editor:
John Fisher

Marketing Director:
Maura Theriault

Channel Manager:
Fair Huntoon

Marketing Coordinator:
Sarena Douglass

Production Director:
Mary Ellen Black

Production Manager:
Andrew Crouth

Production Editor:
Stacy Masucci

Editorial Assistant:
Mary Ellen Martino

Library of Congress Cataloging-in-Publication Data:

Taylor, David L., 1942-
  Elementary blueprint reading for machinists / David L. Taylor.—5th ed.
    p.cm.
  ISBN 1-4018-6256-X
  1. Blueprints. 2. Machinery—Drawings. I. Title

T379.T39 2003
621.8'022'1—dc21

2003046160

NOTICE TO THE READER

Publisher does not warrant or guarantee any of the products described herein or perform any independent analysis in connection with any of the product information contained herein. Publisher does not assume, and expressly disclaims, any obligation to obtain and include information other than that provided to it by the manufacturer.

The reader is expressly warned to consider and adopt all safety precautions that might be indicated by the activities herein and to avoid all potential hazards. By following the instructions contained herein, the reader willingly assumes all risks in connection with such instructions.

The publisher makes no representation or warranties of any kind, including but not limited to, the warranties of fitness for particular purpose or merchantability, nor are any such representations implied with respect to the material set forth herein, and the publisher takes no responsibility with respect to such material. The publisher shall not be liable for any special, consequential, or exemplary damages resulting, in whole or part, from the readers' use of, or reliance upon, this material.

CONTENTS

PREFACE

The reading and interpretation of industrial drawings requires skill development. ELEMENTARY BLUE-PRINT READING FOR MACHINISTS helps develop these skills and applies them to the machine trades and related areas.

This edition of ELEMENTARY BLUEPRINT READING FOR MACHINISTS reflects current practices in industry. The material has been organized and expanded to present a sequential learning experience. New information has been added to explain computer-aided design, new dimensioning practices, and assembly drawings. A list of abbreviations commonly used on drawings appears in the Appendix. Also, a section containing a review of fractions and decimals is included at the end of the text.

The text contains twenty-seven units of instruction and a comprehensive review. Each unit consists of instructional material, an industrial print, and assignment questions. Sketching assignments are also included in some of the units.

Instructional Material

New material is presented in a simple, easy to understand way and then applied to a print. Common shop terminology and applications are described. The related information serves as an aid to reading the print supplied in each unit. It may also be used as a resource for future reference.

Industrial Prints

Each unit contains an industrial print which must be read. These prints increase in reading difficulty in successive units of the text. The prints also incorporate elements of the instructional material learned in previous units.

Assignments

The assignment questions test the students' ability to interpret the blueprint. The various lines, views, dimensions, and notes that make up a blueprint must be understood. Each question examines the level of understanding achieved by the student after the instructional presentation.

Many of the questions require development of mathematical skills. The various dimensional location of lines, surfaces, holes, and other parts must be calculated. The student should review the section on fractions and decimals to answer these questions.

Sketching Assignments

Freehand sketches are quite common in machine trades work. The first step in the development of a design usually is a sketch of the desired part.

Some of the units include an assignment that requires the student to sketch an object. This provides an opportunity to apply principles learned in the instructional section of the unit. It also helps develop skill in sketching without the use of instruments.

Acknowledgments

Robert Brown, Central Carolina Community College, Sanford, NC

Sam Macon, Savannah Technical Institute, Savannah, GA

Danny Moseley, Owensboro Technical College, Owensboro, KY

About the Author

David L. Taylor is currently an Assistant Principal at Bloomington High School North in Bloomington, Indiana. In addition to being a school administrator, Mr. Taylor is a former Journeyman Tool and Die Maker with more than twenty years' experience in vocational–technical training. He holds a Master of Science degree in Adult Education from Penn State University and a Bachelor of Science degree in Vocational–Technical Education from the State University of New York at Buffalo. Mr. Taylor has taught courses in machine trades, print reading, and design at Erie County BOCES, Lewis County BOCES, Jamestown Community College, and Ivy Tech State College. Mr. Taylor is the author of four blueprint reading textbooks published by Delmar Learning.

Industrial Drawings

INTRODUCTION

One of the oldest forms of communication between people is the use of a drawing. A *drawing* is a means of providing information about the size, shape, or location of an object. It is a graphic representation that is used to transfer this information from one person to another.

Drawings play a major role in modern industry. They are used as a highly specialized language among engineers, designers, and others in the technical field. These industrial drawings are known by many names. They are called mechanical drawings, engineering drawings, technical drawings, or working drawings. Whatever the term, their intent remains the same. They provide enough detailed information so that the object may be constructed.

Engineers, designers, and drafting technicians commonly produce drawings using computer-aided design and drafting equipment (CAD). The application of computer technology has led to greater efficiency in drawing production and duplication. CAD systems have rapidly replaced the use of mechanical tools to produce original drawings.

COMPUTER-AIDED DESIGN AND DRAFTING

Computer-aided design or *computer-aided drafting* (CAD) systems are capable of automating many repetitive, time-consuming drawing tasks. The present technology enables the drafter to produce or reproduce drawings to any given size or view. Three-dimensional qualities may also be given to a part, thus reducing the confusion about the true size and shape of an object. Figure 1.1 shows a typical drawing produced with the help of a computer-aided design system.

CAD systems usually consist of three basic components: (1) hardware, (2) software, and (3) operators or users. The hardware includes a processor, a display system, keyboard, plotter, and digitizer often called a "mouse." Software includes the programs required to perform the design or drafting function. Software packages are available in many forms, depending upon the requirements of the user.

The CAD processor is actually the computer or "brains" of the system. The keyboard, which looks very much like a typewriter, is used to place commands into the processor. The commands or input are then displayed graphically on the system display screen. This screen is commonly a cathode ray tube (CRT). The digitizer, or mouse, is used to create graphic images for display on the CRT. The plotter is a printer that produces hard copies of a design in print form.

Industrial drawings are usually produced on a paper material called vellum or on a polyester film material known as Mylar. Mylar is a clear polyester sheet that has a matte finish on one or both sides. The matting provides a dull, granular drawing surface well-suited for pencil or ink lines. Mylar is preferred over vellum in some applications because it resists bending, cracking, and tearing. A completed industrial drawing is known as an original or master drawing.

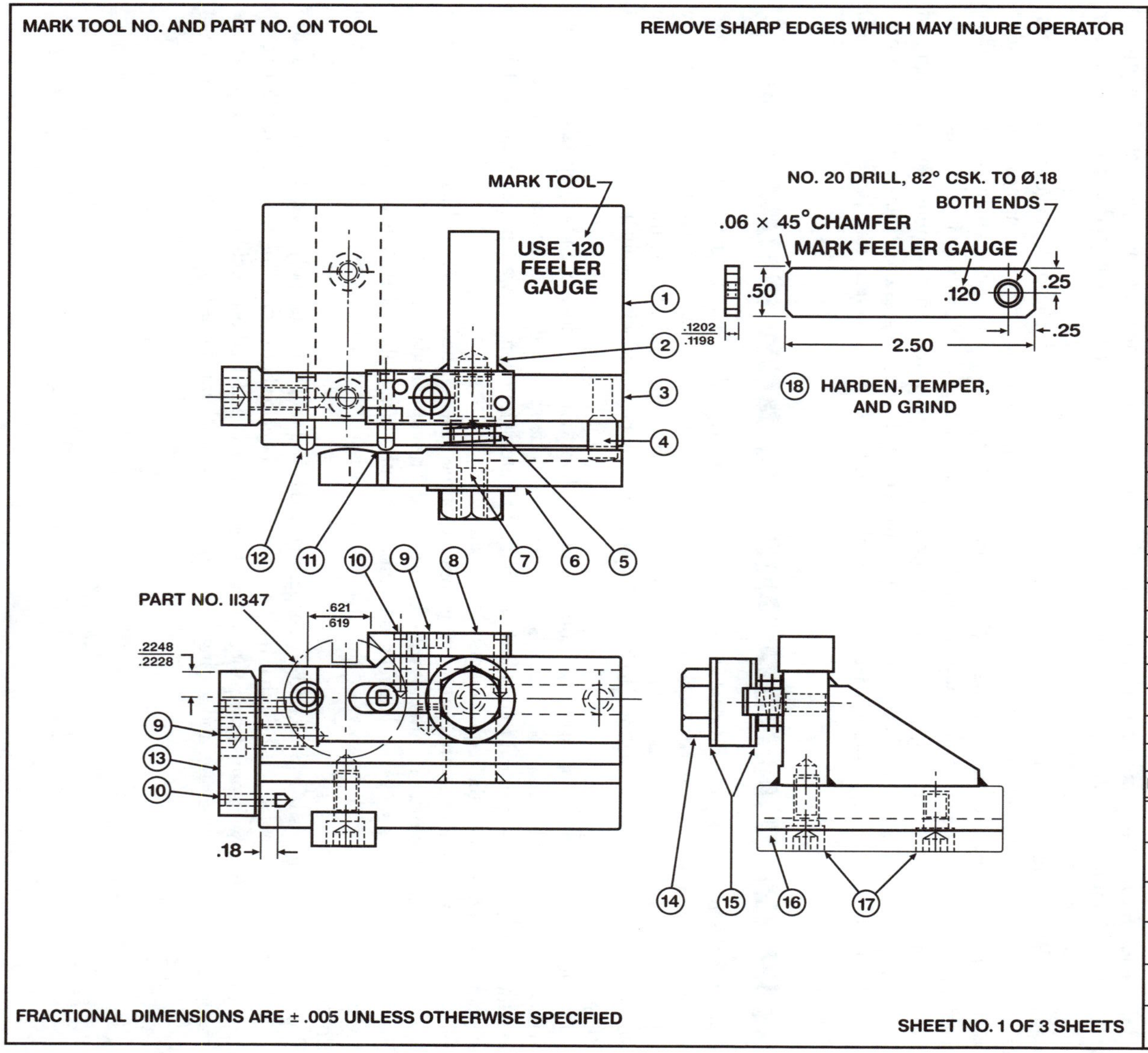

MARK TOOL NO. AND PART NO. ON TOOL

REMOVE SHARP EDGES WHICH MAY INJURE OPERATOR

NO. 20 DRILL, 82° CSK. TO Ø.18
BOTH ENDS

MARK FEELER GAUGE

.06 × 45° CHAMFER

.25

.25

.120

2.50

.50

.1202
.1198

18 HARDEN, TEMPER, AND GRIND

MARK TOOL

USE .120 FEELER GAUGE

1
2
3
4

5
6
7
8
9
10
11
12

PART NO. II347

.2248
.2228

.621
.619

B

.18

9
13
10

17
16
15
14

MATERIAL LIST

DET	QTY	SIZES	MATL
1	1	$\frac{1}{2} \times 2\frac{1}{2} \times 3\frac{3}{4}$	SAE 1020
2	1	$\frac{1}{2} \times 1\frac{1}{2} \times 1\frac{1}{8}$	SAE 1020
3	1	$\frac{1}{2} \times 1\frac{1}{4} \times 3\frac{3}{4}$	SAE 1020
4	1	Ø$\frac{5}{16} \times \frac{15}{16}$ LONG	DR.RD.
5	1	TO SUIT Ø.020	SP.W.
6	1	$\frac{3}{8} \times \frac{7}{8} \times 3\frac{1}{8}$	SAE 1020
7	1	Ø$\frac{7}{16} \times 1\frac{9}{16}$ LONG	SAE 1020
8	1	$\frac{1}{2} \times \frac{5}{8} \times 1\frac{9}{16}$	SAE 1095
9	2	$\frac{1}{4} - 20 \times \frac{1}{2}$ SOCKET HD. CAP SCR.	STD.
10	4	$\frac{1}{8}$ DOWEL $\frac{1}{2}$ LONG	STD.
11	1	Ø$\frac{5}{16} \times 1\frac{3}{8}$ LONG	DR.RD.
12	1	Ø$\frac{5}{16} \times 1\frac{3}{8}$ LONG	DR.RD.
13	1	$\frac{1}{2} \times \frac{5}{8} \times 1\frac{9}{16}$	SAE 1095
14	1	$\frac{5}{16}$ -18 NUT FIN. HEX.	STD.
15	2	$\frac{5}{16}$ W.I. WASHER	STD.
16	1	$\frac{5}{16} \times \frac{5}{8} \times 2\frac{9}{16}$	SAE 1020
17	2	$\frac{1}{4} - 20 \times \frac{3}{8}$ SOCKET HD. CAP SCR.	STD.
18	1	$\frac{1}{8} \times 1\frac{1}{2} \times 2\frac{9}{16}$ GROUND STOCK	B&S

LET'R	DATE	CHANGE	AP'D.
E			
D			
C			
B			
A			

TOOL DRAWING

UNIT	TYPE D5 FOLDING MACHINE
PART NAME	INDEX LOCKING PLATE
PART NO.	11347
TOOL NAME	MILLING FIXTURE
TOOL NO.	T-21463

DRAWN BY	CJO
CHECKED BY	D.B.C.
APROVED BY	L.M.

STAGE NO. 5

DATE 4 -3 -71

SHEET NO. 1 OF 3 SHEETS

FRACTIONAL DIMENSIONS ARE ± .005 UNLESS OTHERWISE SPECIFIED

FIGURE 1.1 ■ Example of an assembly drawing

BLUEPRINTS

Because original drawings are delicate, they seldom leave the drafting room. They are carefully handled and filed in a master file of originals. When a copy of an original is required, a print is made.

The term used for the process of reproducing an original is known as *blueprinting*. The earliest form of blueprinting produced white line, blue background reproductions. This early process, which was developed in England more than 100 years ago, has since changed. Modern reproductions produce a dark line, white background duplication simply called a *print*. However, the term blueprint is still widely used in industry and has been included in the title of this text.

INTERPRETING INDUSTRIAL DRAWINGS

Industrial drawings and prints are made for the purpose of communication. They are a form of nonverbal communication between a designer and builder of a product. Industrial drawings are referred to as a universal language. It is a language that can be interpreted and understood regardless of country. Also, drawings and prints become part of a contract between parties buying and selling manufactured parts.

A picture or photograph of an object would show how the object appears. However, it would not show the exact size, shape, and location of the various parts of the object.

Industrial drawings describe size and shape and give other information needed to construct the object. This information is presented in the form of special lines, views, dimensions, notes, and symbols. The interpretation of these elements is called *print reading*.

THE REPRODUCTION PROCESSES

There are several methods available for reproducing drawings.

Chemical Process

The ammonia process is a common method of print reproduction. To produce a copy, the original is placed on top of a light-sensitive print paper. Both the original and the print are fed into the diazo machine and exposed to a strong ultraviolet light. As the light passes through the thin original, it burns off all sensitized areas not shadowed by lines. The print paper is then exposed to an ammonia atmosphere. The ammonia develops all sensitized areas left on the print paper. The result is a dark line reproduction on a light background.

Silver Process

The silver process is actually a photographic method of reproduction. This process is often referred to as microfilming or photocopying.

This method is rapidly gaining popularity in industry due to storage and security reasons. The most common procedure followed is to photograph an original drawing to gain a microfilm negative. The negative is then placed on an aperture card and labeled with a print number. Duplicates of the microfilm are produced with the aid of a microfilm printer using sensitized photographic materials. Enlarged or reduced prints can be produced using this process.

The aperture cards containing the microfilm are very small. Therefore, cataloging and filing take very little room for storage. They are also much easier to handle than the delicate originals, which must be kept in large files.

Microfilming is often done for security reasons. As many as 200 prints may be placed on one roll of microfilm. They may then be placed in a vault or other secure area.

Electrostatic Process

The electrostatic process has gained in popularity for industrial drawing reproduction. Although once limited to reproducing documents and small drawings, new machines have been developed that allow large drawing duplication. The electrostatic process, commonly known as xerography, uses a zinc-coated paper that is given an electrostatic charge. The zinc coating is sensitive to ultraviolet light when exposed. Areas shadowed by lines on the original produce a dark line copy.

CAD Process

One advantage of a CAD system is the ability to file and store original drawings electronically. Stored drawings can be accessed and reproduced whenever a revision is required or additional copies are needed. To reproduce a CAD drawing, a message must be sent from the CAD processor to an output device called a printer or printer/plotter, Figure 1.2.

FIGURE 1.2 ■ CAD printer/plotter

ASSIGNMENT: REVIEW QUESTIONS

1. List two other names commonly given industrial drawings.

    a. _____

    b. _____

2. Industrial drawings should provide enough information so that the object can be _____.

3. The paper material on which original drawings are produced is called _____.

4. A completed industrial drawing is known as a master drawing or _____.

5. Master drawings

    a. are provided to the machine builder.

    b. seldom leave the drafting room.

    c. are developed by the master drafter.

    d. are always drawn on vellum.

6. What is the term used for reproducing an industrial drawing?

    _____

    _____

7. Industrial drawings are often referred to as _____ language.

8. Industrial drawings are a form of communication that is

    a. verbal.

    b. nonverbal.

9. Why is a photograph not used to describe an object?

    _____

    _____

10. The light the print paper is exposed to in the diazo process is

    a. sunlight.

    b. infrared light.

    c. fluorescent light.

    d. ultraviolet light.

11. The silver process is

    a. seldom used.

    b. a photographic process.

    c. an ammonia process.

    d. a heat process.

12. List two advantages of microfilming.

    a. _____

    b. _____

13. Aperture cards

    a. are small.

    b. contain print information.

    c. are used for filing.

    d. all of the above.

    e. none of the above.

14. The heat process uses a chemically coated paper that is sensitive to
    a. infrared light.
    b. heat.
    c. ammonia.
    d. ultraviolet light.

15. The electrostatic process uses paper that is sensitive to
    a. chemicals.
    b. ammonia.
    c. heat.
    d. light.

16. The electrostatic process uses a paper coated with
    a. carbon.
    b. lead.
    c. iron.
    d. zinc.

17. List three components of a CAD system.
    a. _____
    b. _____
    c. _____

18. The display screen used with a CAD system is called a _____.

19. What is one advantage a CAD system has over conventional drawing methods?

    _____

    _____

UNIT 2
Title Blocks

All industrial drawings have certain elements in common. They consist of various lines, views, dimensions, and notes. Other general information is also supplied so that the object may be completely understood. The skilled print reader must learn to interpret and apply the information provided on the drawing.

TITLE BLOCKS

A *title block* or *title strip* is designed to provide general information about the part, assembly, or the drawing itself. Title blocks are usually located in the lower right-hand corner of the print, Figure 2.1. Title strips extend along the entire lower section of the print, Figure 2.2. The location of each depends on the filing system each company uses.

Most companies select a standard title form for their drawings that is printed on the original drafting sheet, Figure 2.3. This enables the drafter to simply fill in the required information.

The most common information found in the title block or strip includes the following:

■ *Company name* identifies the company using or purchasing the drawing.

■ *Part name* identifies the part or assembly drawn.

■ *Part number* identifies the number of the part for manufacturing or purchasing information.

■ *Drawing number* is used for reference when filing the original drawing.

■ *Scale* indicates the relationship between the size of the drawing and the actual size of the part. This scale may be a full-size scale of 1 = 1; half-size scale of $\frac{1}{2}$ = 1 or 6 inches on drawing equals 12 inches on the part; quarter-size scale of $\frac{1}{4}$ = 1 or 3 inches equals 12 inches; etc.

			CUST. _____
			CITY _____
			C.O. _____ S.Q. _____
			QUAN. DATE
RING DIVISION	DR. JSP DATE 4/18/95	**AISI-** 02 **Rc-** 60 – 63	REFERENCE
PRODUCTO MACHINE CO. JAMESTOWN, NEW YORK 14701	CK. DATE		
	VG NO. —		
DWG. 24934–2 REV.	TEMP NO. T –		LATEST CHANGE / REC'D

FIGURE 2.1 ■ Sample title block

					DATE	DWN BY:	CKD BY:	APPR. BY:
					4/25/95	DLT	JLS	TRC
1	1.250 WAS 1.000	5-2-95	AWT		SCALE:		MATERIAL:	
NO.	CHANGE	DATE	BY		FULL		SAE 2335	
STANDARD TOLERANCES UNLESS OTHERWISE SPECIFIED					PART NAME: CONTROL BRACKET			⊕◁
FRACTIONAL ± 1/64 2 PLC. DECIMAL ± .01 3 PLC. DECIMAL ± .005 4 PLC. DECIMAL ± .0005 LIMITS ON ANGULAR DIMENSIONS ± 1/2° FINISH: BREAK ALL SHARP CORNERS					PART NUMBER: A01-3002424-005			D-15

FIGURE 2.2 ■ Sample title strip

STANDARD TOLERANCE UNLESS OTHERWISE SPECIFIED		DET.	SHT.	DESCRIPTION		STOCK: FIN. ALLOWED	MAT.	HT. TR.	REQ'D
		BILL OF MATERIAL ONE							
SPREAD BETWEEN SCREW HOLES MUST BE HELD TO A TOLERANCE OF ±.008 AND SPREAD BETWEEN DOWEL HOLES MUST BE HELD TO A TOLERANCE OF ±.0005		**ABC MACHINE COMPANY**							
MILLIMETER	INCH							JAMESTOWN, NEW YORK	
		TOOL NAME							
WHOLE NO. ± 0.5 1 PLC. DEC ± 0.2 2 PLC. DEC ± 0.03 3 PLC. DEC ± 0.013	FRACTIONAL ± 1/64 2 PLC. DEC ± 0.01 3 PLC. DEC ± 0.001 4 PLC. DEC ± 0.0005	FOR:							
		OPER:							
		MACHINE:					DATE		
ANGLE ± 1/2°		DR.		SCALE	PART No.				
BREAK ALL SHARP CORNERS AND EDGES UNLESS OTHERWISE SPECIFIED		CH.		No. OF SHEETS					
		APP.		SHEET No.	TOOL No.				

FIGURE 2.3 ■ An example of an industrial title block

■ *Tolerance* refers to the amount that a dimension may vary from the print. Standard tolerances that apply to the entire print are given in the title block. Tolerances referring to only one surface are indicated near that surface on the print.

■ *Material* indicates the type of material of which the part is to be made.

■ *Heat treat information* provides information as to hardness or other heat treat specifications.

■ *Date* identifies the date the drawing was made.

■ *Drafter* identifies who prepared the original.

■ *Checker* identifies who checked the completed drawing.

■ *Approval* identifies who approved the design of the object.

■ *Change notes or revision* is an area in the block that records for history changes that are made on the drawing. Often revision blocks are located elsewhere on the drawing.

STANDARD ABBREVIATIONS FOR MATERIALS

A variety of materials are used in industry. The drafter or designer must select materials that will best fit the job application. The ability to do this comes from experience and from understanding material characteristics.

To save time and drawing space, material specifications are usually abbreviated on drawings. Table 2–1 describes the most common abbreviations used. Refer to this table as a guide to material abbreviations used later in the text. Additional tables are found in the Appendix.

TABLE 2–1 STANDARD ABBREVIATIONS FOR MATERIALS			
Alloy Steel	AL STL	Hot-Rolled Steel	HRS
Aluminum	AL	Low-Carbon Steel	LCS
Brass	BRS	Magnesium	MAG
Bronze	BRZ	Malleable Iron	MI
Cast Iron	CI	Nickel Steel	NS
Cold-Drawn Steel	CDS	Stainless Steel	SST
Cold-Finished Steel	CFS	Steel	STL
Cold-Rolled Steel	CRS	Tool Steel	TS
High-Carbon Steel	HCS	Tungsten	TU
High-Speed Steel	HSS	Wrought Iron	WI

PARTS LISTS

A *parts list,* also called a *bill of materials,* is often included with the blueprint, Figure 2.4. This list provides information about all parts required for a complete assembly of individual details. The bill of materials is most frequently found on the print that displays the completed assembly and is known as the *assembly drawing.* The assembly drawing is a pictorial representation of a fully assembled unit that has all parts in their working positions.

Additional drawings called *detail drawings* usually accompany the assembly drawing and are numbered for identification. Each assembly detail found in the bill of materials is also provided with a reference number that is used to locate the detail on the detail drawing. Detail drawings give more complete information about the individual units.

Assembly drawings are covered more completely in a later unit of the text.

ASSIGNMENT D-1: RADIUS GAUGE

1. What is the name of the part? _____

2. What is the part number? _____

3. What is the scale of the drawing? _____

4. Of what material is the part made? _____

5. What finish is required? _____

6. What tolerances are allowed on two-place decimal dimensions? _____

7. What are the tolerances allowed on three-place decimal dimensions? _____

8. What are the tolerances allowed on the fractional dimensions? _____

9. What are the tolerances allowed on the angular dimensions? _____

10. What is another name for the parts list? _____

11. What is the area on the drawing where general information is provided? _____

12. What is the number used for filing drawings called? _____

13. Have any changes been indicated on the radius gauge? _____

14. What are copies of originals called? _____

15. What is the date of this drawing? _____

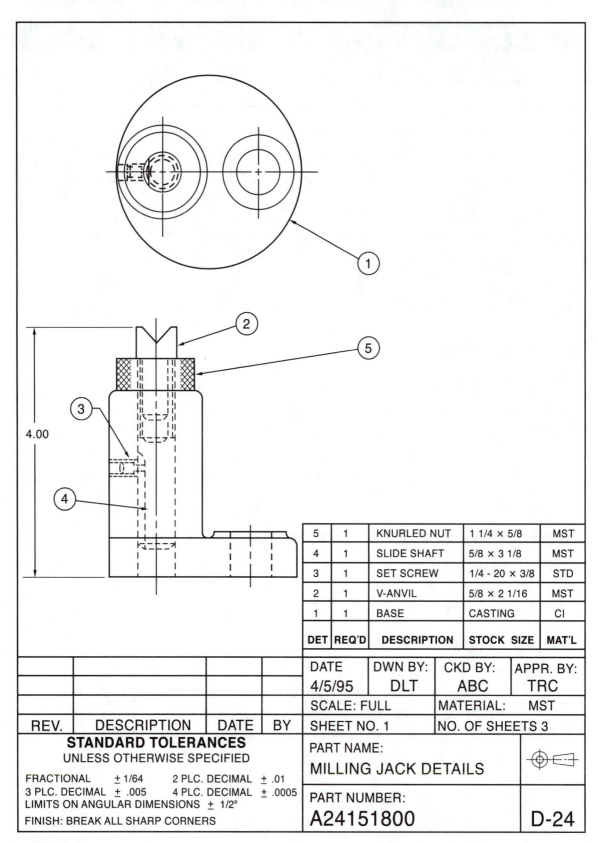

5	1	KNURLED NUT	1 1/4 × 5/8	MST
4	1	SLIDE SHAFT	5/8 × 3 1/8	MST
3	1	SET SCREW	1/4 - 20 × 3/8	STD
2	1	V-ANVIL	5/8 × 2 1/16	MST
1	1	BASE	CASTING	CI
DET	**REQ'D**	**DESCRIPTION**	**STOCK SIZE**	**MAT'L**

				DATE	DWN BY:	CKD BY:	APPR. BY:
				4/5/95	DLT	ABC	TRC
				SCALE: FULL		MATERIAL:	MST
REV.	DESCRIPTION	DATE	BY	SHEET NO. 1		NO. OF SHEETS 3	

STANDARD TOLERANCES
UNLESS OTHERWISE SPECIFIED

FRACTIONAL ± 1/64   2 PLC. DECIMAL ± .01
3 PLC. DECIMAL ± .005   4 PLC. DECIMAL ± .0005
LIMITS ON ANGULAR DIMENSIONS ± 1/2°
FINISH: BREAK ALL SHARP CORNERS

PART NAME:
MILLING JACK DETAILS

PART NUMBER:
A24151800

D-24

FIGURE 2.4 ■ Example of a parts list on an assembly drawing

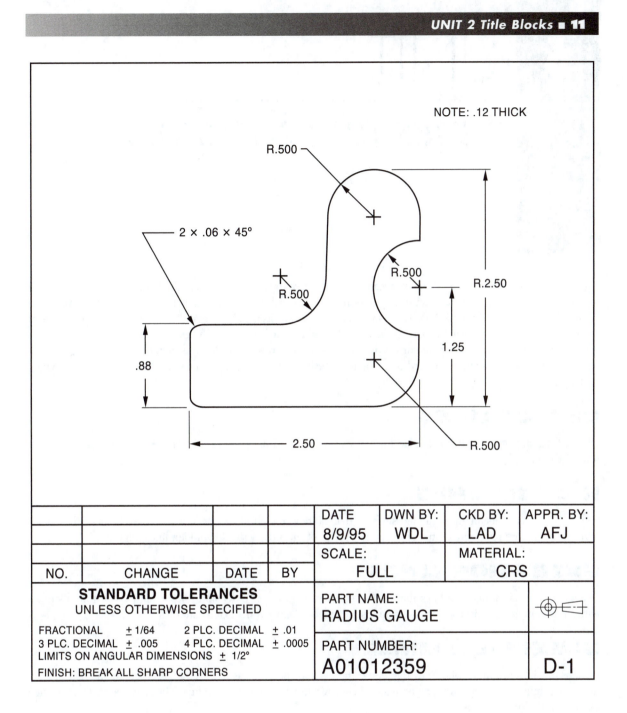

NOTE: .12 THICK

R.500

2 × .06 × 45°

R.500

R.500

R.2.50

1.25

.88

2.50

R.500

NO.	CHANGE	DATE	BY	DATE 8/9/95	DWN BY: WDL	CKD BY: LAD	APPR. BY: AFJ
				SCALE: FULL		MATERIAL: CRS	

STANDARD TOLERANCES
UNLESS OTHERWISE SPECIFIED

FRACTIONAL ± 1/64   2 PLC. DECIMAL ± .01
3 PLC. DECIMAL ± .005   4 PLC. DECIMAL ± .0005
LIMITS ON ANGULAR DIMENSIONS ± 1/2°
FINISH: BREAK ALL SHARP CORNERS

PART NAME:
RADIUS GAUGE

PART NUMBER:
A01012359

D-1

3 UNIT
Lines and Symbols

Various lines on a drawing have different meanings. They may appear solid, broken, thick, or thin. Each is designed to help the blueprint reader make an interpretation. The standards for these lines were developed by the American National Standards Institute (ANSI). These lines are now known as the alphabet of lines, Figure 3.1. Knowledge of these lines helps one visualize the part. Some lines show shape, size, centers of holes, or the inside of a part. Others show dimensions, positions of parts, or simply aid the drafter in placing the various views on the drawing.

This unit describes the most basic lines. The identification of other types of lines will be described in following units.

OBJECT LINES

Object lines are heavy, solid lines also known as *visible edge lines,* Figure 3.2. They generally show the outline of the part.

HIDDEN LINES

Some objects have one or more hidden surfaces that cannot be seen in the given view. These hidden surfaces, or invisible edges, are represented on a drawing by a series of short dashes called hidden lines, Figure 3.3.

EXTENSION LINES

Extension lines are thin, solid lines that extend surfaces, Figure 3.4. Extension lines extend away from a surface without touching the object. Dimensions are usually placed between the extension lines.

DIMENSION LINES

Dimension lines are thin, solid lines that show the distance being measured, see Figure 3.4. At the end of each dimension line is an *arrowhead*. The points of the arrows touch each extension line. The space being dimensioned extends to the tip of each arrow.

Arrowheads may be open or solid and can vary in size. The size depends mostly on the dimension line weight and blueprint size.

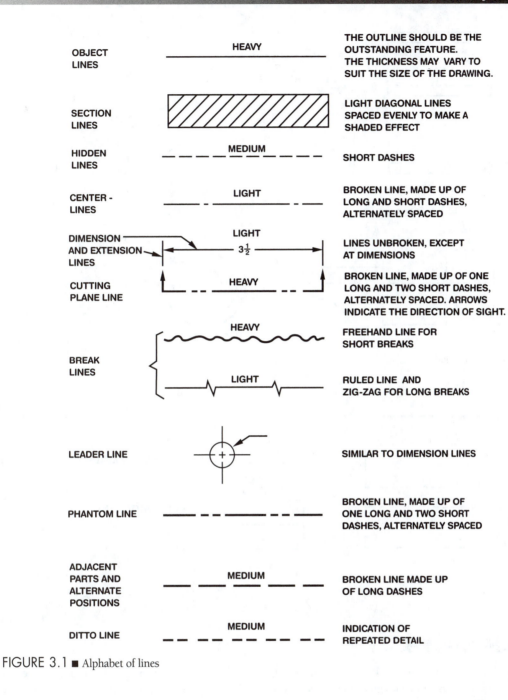

OBJECT LINES	HEAVY	THE OUTLINE SHOULD BE THE OUTSTANDING FEATURE. THE THICKNESS MAY VARY TO SUIT THE SIZE OF THE DRAWING.
SECTION LINES		LIGHT DIAGONAL LINES SPACED EVENLY TO MAKE A SHADED EFFECT
HIDDEN LINES	MEDIUM	SHORT DASHES
CENTER - LINES	LIGHT	BROKEN LINE, MADE UP OF LONG AND SHORT DASHES, ALTERNATELY SPACED
DIMENSION AND EXTENSION LINES	LIGHT $3\frac{1}{2}$	LINES UNBROKEN, EXCEPT AT DIMENSIONS
CUTTING PLANE LINE	HEAVY	BROKEN LINE, MADE UP OF ONE LONG AND TWO SHORT DASHES, ALTERNATELY SPACED. ARROWS INDICATE THE DIRECTION OF SIGHT.
BREAK LINES	HEAVY	FREEHAND LINE FOR SHORT BREAKS
	LIGHT	RULED LINE AND ZIG-ZAG FOR LONG BREAKS
LEADER LINE		SIMILAR TO DIMENSION LINES
PHANTOM LINE		BROKEN LINE, MADE UP OF ONE LONG AND TWO SHORT DASHES, ALTERNATELY SPACED
ADJACENT PARTS AND ALTERNATE POSITIONS	MEDIUM	BROKEN LINE MADE UP OF LONG DASHES
DITTO LINE	MEDIUM	INDICATION OF REPEATED DETAIL

FIGURE 3.1 ■ Alphabet of lines

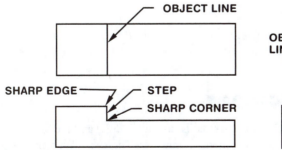
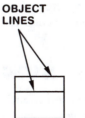
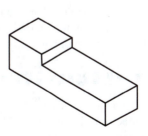

FIGURE 3.2 ■ Object or visible edge lines

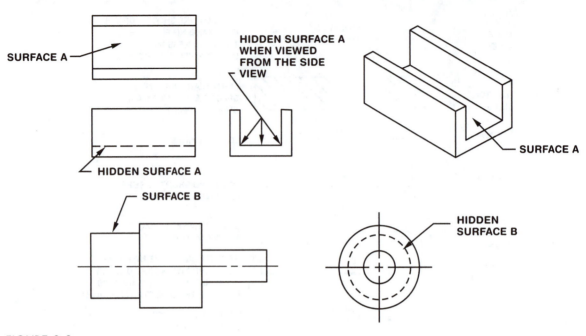

FIGURE 3.3 ■ Hidden surfaces

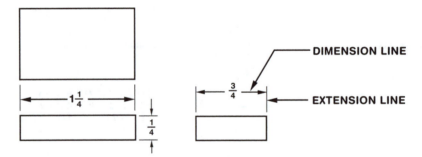

FIGURE 3.4 ■ Extension and dimension lines

CENTERLINES

Centerlines are thin lines with alternate long and short dashes. They do not form part of the object, but are used to show a location. As the name implies, centerlines indicate centers. They are used to show centers of circles, arcs, or symmetrical parts, Figure 3.5.

LEADER LINES

Leader lines are similar in appearance to dimension lines. They consist of an inclined line with an arrow at the end where the dimension or surface is being called out. The inclined line is attached to a horizontal leg at the end of which a dimension or note is provided, Figure 3.6.

APPLICATION OF SYMBOLS

Revised drawing standards developed by the American National Standards Institute (ANSI) and the American Society of Mechanical Engineers (ASME) are being applied to most modern drawings. These standards encourage the use of symbols to replace words or notes on drawings. This practice reduces drafting time, reduces the amount of written information on the drawing, and helps overcome language barriers. Figure 3.7 shows some common symbols applied to prints. The application of most of these symbols is explained in the appropriate units that follow.

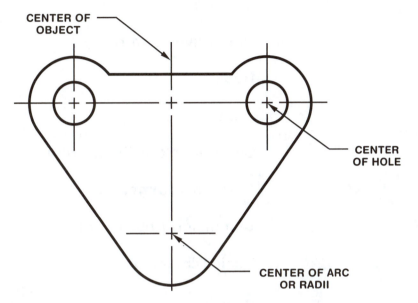

FIGURE 3.5 ■ Centerlines

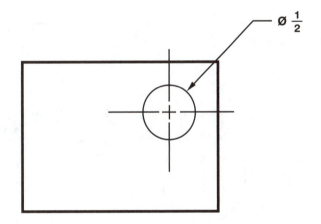

FIGURE 3.6 ■ Leader lines

DIAMETER SYMBOLS

The former practice was to specify holes or diameters by calling out the hole size, using an abbreviation or letter for the diameter, DIA or D, and a note for the process, Figure 3.8.

The new standard for diameter uses the symbol $\varnothing$ in front of the dimension indicating a diameter and the reference to a machining process is not given, Figure 3.9.

However, industrial use of the latest standards varies. Many drawings still reflect the older methods of dimensioning.

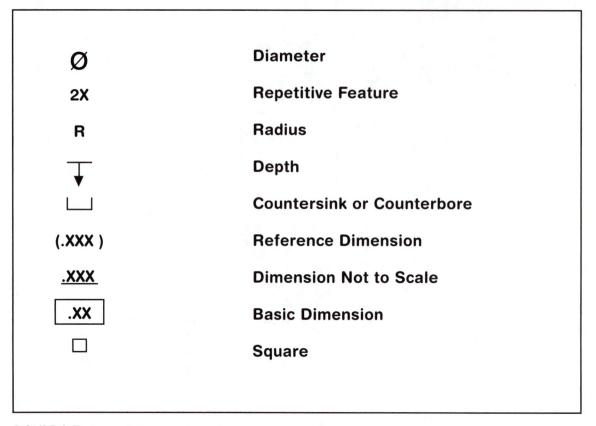

Ø	**Diameter**
2X	**Repetitive Feature**
R	**Radius**
↧	**Depth**
⌴	**Countersink or Counterbore**
(.XXX)	**Reference Dimension**
.XXX	**Dimension Not to Scale**
.XX	**Basic Dimension**
□	**Square**

FIGURE 3.7 ■ Standard feature symbols

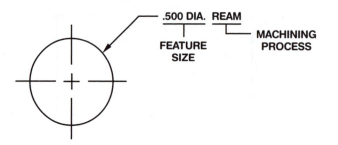

FIGURE 3.8 ■ Old method of specifying a diameter and process

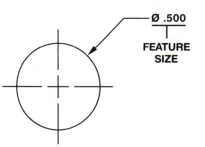

FIGURE 3.9 ■ New method of specifying a diameter

SQUARE SYMBOL

A square symbol is often used to show that a single dimension applies to a square shape. The use of a square symbol preceding a dimension indicates that the feature being called out is square, Figure 3.10.

SPECIFYING REPETITIVE FEATURES

Repetitive features or dimensions are often specified in more than one place on a drawing. To eliminate the need of dimensioning each individual feature, notes or symbols may be added to show that a process or dimension is repeated.

Holes of equal size may be called out by specifying the number of features required by an X. A small space is left between the X and the feature size dimension that follows, Figure 3.11.

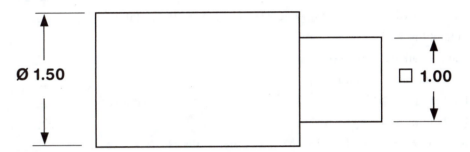

FIGURE 3.10 ■ Application of a square symbol to represent a feature

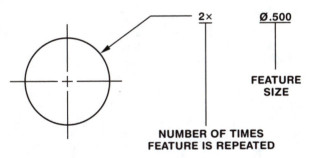

FIGURE 3.11 ■ New method of representing repetitive features

ASSIGNMENT D-2: TOP PLATE

1. What is the name of the part?   _____

2. What is the part number?   _____

3. Of what material is the part made?   _____

4. How thick is the part?   _____

5. What kind of line is Ⓐ?   _____

6. What radius forms the front of the plate?   _____

7. How many holes are there?   _____

8. What kind of line is Ⓑ?   _____

9. How far are the centers of the two holes from the vertical centerline of the piece?   _____

10. How far apart are the centers of the two holes?   _____

11. What radius is used to form the two large diameters around the 1.000 holes?   _____

12. What kind of line is Ⓒ?   _____

13. What diameter are the two holes?   _____

14. What does the symbol 2X mean?   _____

15. What kind of line is Ⓓ?   _____

16. What kind of line is Ⓔ?   _____

17. What is the overall distance from left to right of the top plate?   _____

18. What kind of a line is drawn through the center of a hole?   _____

19. What is the scale of the drawing?   _____

20. What special finish is required on the part?   _____

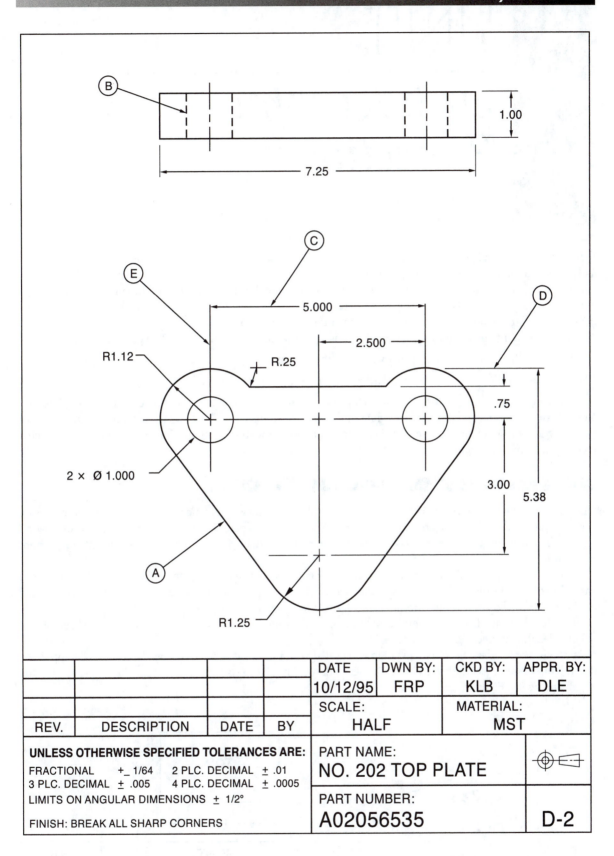

				DATE	DWN BY:	CKD BY:	APPR. BY:
				10/12/95	FRP	KLB	DLE
				SCALE:		MATERIAL:	
REV.	DESCRIPTION	DATE	BY	HALF		MST	

UNLESS OTHERWISE SPECIFIED TOLERANCES ARE:	PART NAME:	
FRACTIONAL +_ 1/64 2 PLC. DECIMAL ± .01 3 PLC. DECIMAL ± .005 4 PLC. DECIMAL ± .0005 LIMITS ON ANGULAR DIMENSIONS ± 1/2°	NO. 202 TOP PLATE	
FINISH: BREAK ALL SHARP CORNERS	PART NUMBER: A02056535	D-2

4 UNIT

Orthographic Projection

Industrial drawings and prints furnish a description of the shape and size of an object. All information necessary for its manufacture must be presented in a form that is easily recognized. For this reason, a number of views are necessary. Each view shows a part of the object as it is seen by looking directly at each one of the surfaces. When all the notes, symbols, and dimensions are added to the projected views, it becomes a working drawing. A *working drawing* supplies all the information required to construct the part, Figure 4.1.

The ability to interpret a drawing accurately is based on the mastery of two skills. The print reader must:

1. Visualize the completed object by examining the drawing itself.

2. Know and understand certain standardized signs and symbols.

Visualizing is the process of forming a mental picture of an object. It is the secret of successful drawing interpretation. Visualization requires an understanding of the exact relationship of the views to each other. It also requires a working knowledge of how the individual views are obtained through projection. When these views are connected mentally, the object has length, width, and thickness.

PRINCIPLES OF PROJECTION

Most objects can be drawn by projecting them onto the sheet in some combination of the front, top, and right-side views. To project the views of an object into the three views, imagine it placed in a square box with transparent sides, Figure 4.2. The top is hinged to swing directly over the front. The right side is hinged to swing directly to the right of the front.

In this case, the surfaces of the object selected are rectangular in shape. The front surface of the object is placed parallel to the front surface of the box. With the object held in this position, the outline of its front surface is traced on the face of the box as it would appear to the observer looking directly at it.

Note that the front of the object, indicated by A, B, C, D as drawn on the front surface in its correct shape, shows only the length and thickness, Figure 4.3.

Without moving the object, the operation is repeated with the observer looking directly down at the top. Note that the top of the object, indicated by A, D, F, G as drawn on the top surface in its correct shape, shows only the length and width, Figure 4.4.

A PICTURE

A WORKING DRAWING

FIGURE 4.1 ■ Projecting an object into three views

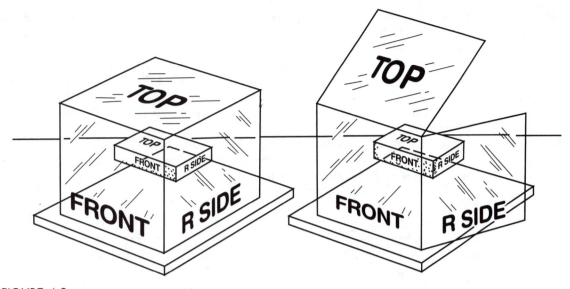

FIGURE 4.2 ■ Box with transparent sides

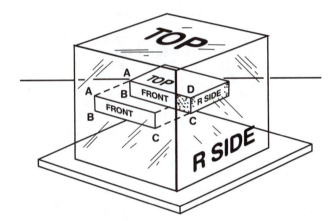

FIGURE 4.3 ■ Projecting the front view

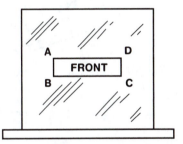

FRONT VIEW AS SEEN DIRECTLY FROM THE FRONT

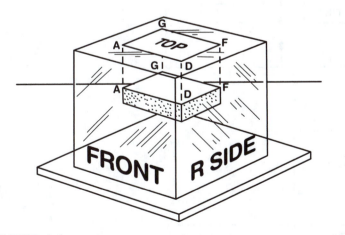

FIGURE 4.4 ■ Projecting the top view

TOP VIEW AS SEEN DIRECTLY FROM ABOVE

The operation is repeated again for the side view with the observer looking directly at the right side. Note that the right side of the object, indicated by D, C, E, F as drawn on the right side surface in its correct shape, shows only the width and thickness, Figure 4.5.

If the top of the box is swung upward to a vertical position, the top view would appear directly over the front view. If the right side of the box is swung forward, the side view would appear to the right and in line with the front view, Figure 4.6.

The sides of the box and the identification letters are now removed, leaving the three projected views of the object in their correct relation, Figure 4.7.

A drawing has now been made of each of the three principal views (the front, the top, and the right side). Each shows the exact shape and size of the object and the relationship of the three views to each other. This principle is called *orthographic projection* and is used throughout all mechanical drawing.

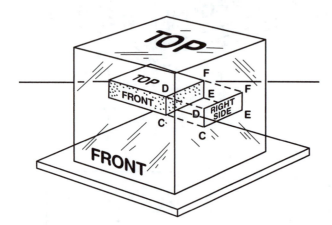

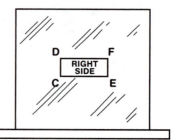

RIGHT-SIDE VIEW AS SEEN DIRECTLY FROM THE RIGHT SIDE

FIGURE 4.5 ■ Projecting the right-side view

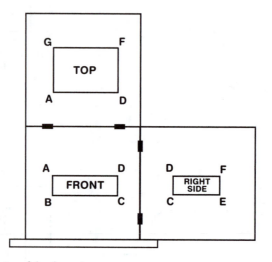

FIGURE 4.6 ■ The correct relation of the three views

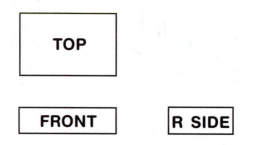

FIGURE 4.7 ■ The projected views with the projection aids removed

ARRANGEMENT OF VIEWS

The three-view drawing illustrated in Figure 4.7 shows the relative positions of the top, front, and right-side views. Often more or fewer views are needed to explain all the details of the part. The shape and the complexity of the object determines the number and arrangement of views. The drafter should supply enough detailed views of information for the construction of the object. Part of a drafter's job is to decide which views will best accomplish this purpose.

Figure 4.8 shows the position of views as they might appear on a working drawing.

The name and location of each view is identified throughout this text as follows:

Front View	F. V.
Top View	T. V.
Right-side View	R. V.
Left-side View	L. V.
Bottom View	Bot. V.
Auxiliary View	Aux. V.
Back or Rear View	B. V.
Isometric View	I. V.

The back view may be located in any one of the places indicated in Figure 4.8.

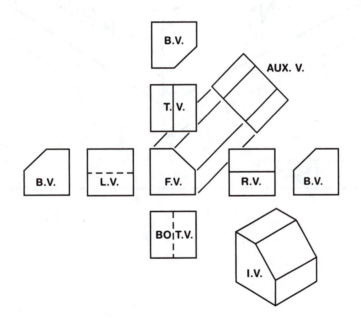

FIGURE 4.8 ■ Relative positions of views

SKETCH S-1: DIE BLOCK

1. Lay out the front, top, and right-side views.

2. Start the sketch about 1/2 inch from the left-hand margin and about 1/2 inch from the bottom. Make the views 1 inch apart.

3. Dimension the completed sketch.

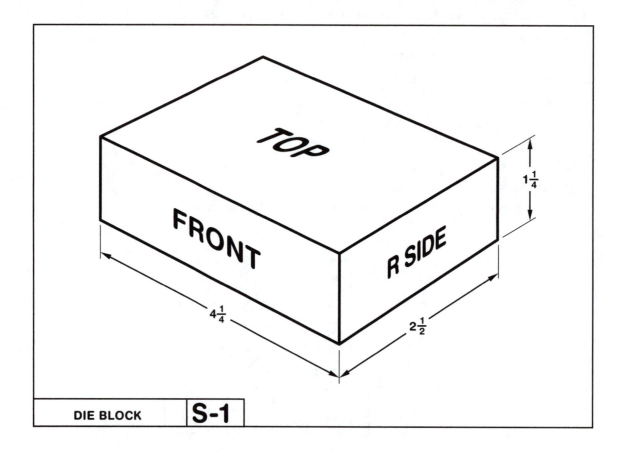

DIE BLOCK | **S-1**

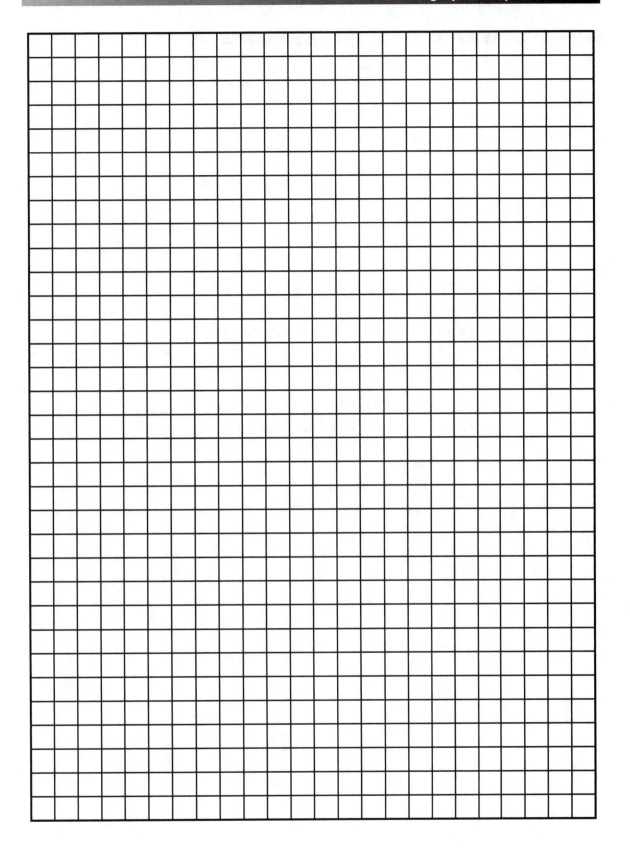

ASSIGNMENT D-3: PRESSURE PAD BLANK

Note: Letters are used on some of these drawings so that questions may be asked about lines and surfaces without involving a great number of descriptive items. They are learning aids and, as the course progresses, are omitted from the more advanced problems.

1. What is the name of the part? _____

2. What is the date of this drawing? _____

3. What is the part number? _____

4. How many views of the blank are shown? (Do not include the sketch.) _____

5. What is the overall length? _____

6. What is the overall height or thickness? _____

7. What is the overall depth or width? _____

8. In what two views is the length the same? _____

9. In what two views is the height, or thickness, the same? _____

10. In what two views is the width the same? _____

11. If Ⓕ represents the top surface, what line in the front view represents the top of the object? _____

12. If Ⓗ represents the surface in the right-side view, what line represents this surface in the top view? _____

13. What line in the top view represents the surface Ⓖ of the front view? _____

14. What line in the right-side view represents the front surface of the front view? _____

15. What surface shown does Ⓙ represent? _____

16. What line of the front view does line Ⓐ represent? _____

17. What line of the top view does point Ⓑ represent? _____

18. What line of the top view does line Ⓒ represent? _____

19. What kind of line is Ⓛ? _____

20. What kind of line is Ⓝ? _____

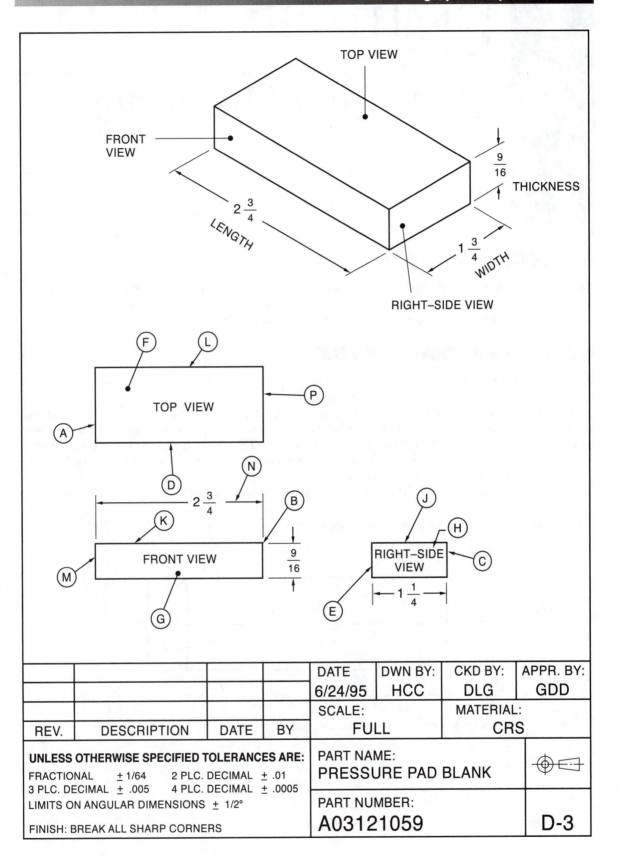

TOP VIEW

FRONT
VIEW

$9\over16$

THICKNESS

$2{3\over4}$
LENGTH

$1{3\over4}$
WIDTH

RIGHT–SIDE VIEW

F   L

TOP VIEW

P

A

D   N

$2{3\over4}$

K

B

$9\over16$

M   FRONT VIEW

G

J

H

RIGHT–SIDE
VIEW

C

E $1{1\over4}$

				DATE	DWN BY:	CKD BY:	APPR. BY:
				6/24/95	HCC	DLG	GDD
				SCALE:		MATERIAL:	
REV.	DESCRIPTION	DATE	BY	FULL		CRS	

UNLESS OTHERWISE SPECIFIED TOLERANCES ARE:	PART NAME:	
FRACTIONAL ± 1/64 2 PLC. DECIMAL ± .01 3 PLC. DECIMAL ± .005 4 PLC. DECIMAL ± .0005 LIMITS ON ANGULAR DIMENSIONS ± 1/2° FINISH: BREAK ALL SHARP CORNERS	PRESSURE PAD BLANK	
	PART NUMBER: A03121059	D-3

5 UNIT
One-View Drawings

SELECTION OF VIEWS

The selection and number of views placed on a drawing are determined by the drafter. A drawing should show the object in as few views as possible for clear and complete shape description. Simple objects may be shown in one view. As the object becomes more complex, more views may be required to describe the object.

ONE-VIEW DRAWINGS

Different views of an object help the print reader visualize the part. Each view gives information describing the size and shape of the object as it is seen from different sides.

On work that is uniform in shape, only one view may be given. This is often the case with cylindrical work such as simple bolts, shafts, pins, or rods. Additional information about the part is provided in the form of notes or symbols. This saves drafting time and also makes the print easier to read because unnecessary views are not shown.

In the case of Figures 5.1 and 5.2, the side or circular views would be omitted. The centerlines and the symbol $\varnothing$ for diameter indicate that the objects are cylindrical.

Parts that are flat and thin may also be drawn using one view. Notes are added to describe thickness, material, or operations, Figure 5.3.

When only one view is drawn, it generally is called a front view.

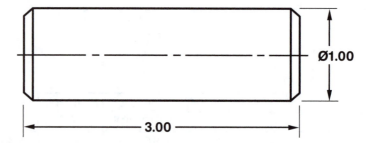

FIGURE 5.1 ■ Only one view is necessary to describe this cylindrical object

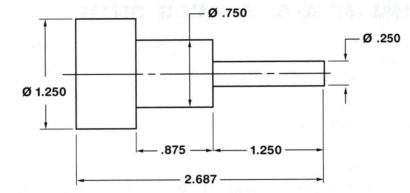

FIGURE 5.2 ■ ■ The centerline and symbols indicate that the object is cylindrical

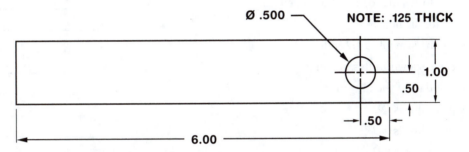

FIGURE 5.3 ■ ■ One view and additional notes give all the necessary information about the object

ASSIGNMENT D-4: SPACER SHIM

1. What is the name of the part? _____

2. What is the scale of the drawing? _____

3. What does the 3 × mean? _____

4. What is the part number? _____

5. Of what material is the part made? _____

6. How thick is the part? _____

7. How many holes are required? _____

8. What is the overall length of the shim from left to right? _____

9. What is the distance between holes? _____

10. What is the radius around the end of the shim? _____

11. What is the overall width from top to bottom? _____

12. What angle forms the top and bottom edges? _____

13. What size are the holes? _____

14. What does the symbol ⌀ 1/2 indicate? _____

15. What view is shown in this drawing? _____

16. What is the date of the drawing? _____

17. What type of line is Ⓐ? _____

18. What type of line is Ⓑ? _____

19. What type of line is Ⓒ? _____

20. What type of line is Ⓓ? _____

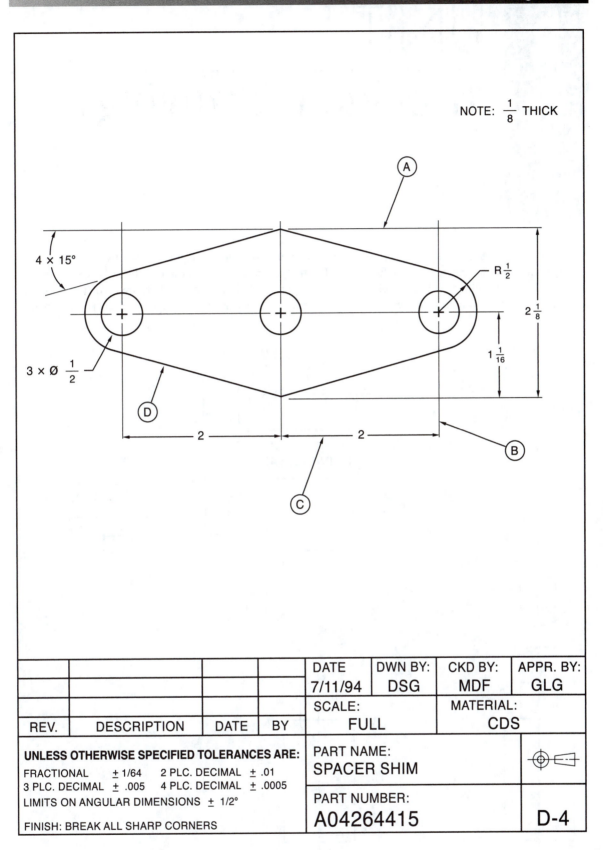

NOTE: $\frac{1}{8}$ THICK

Ⓐ

$4 \times 15°$

$R\frac{1}{2}$

$3 \times \varnothing\ \frac{1}{2}$

$2\frac{1}{8}$

$1\frac{1}{16}$

Ⓓ

2

2

Ⓑ

Ⓒ

DATE	DWN BY:	CKD BY:	APPR. BY:
7/11/94	DSG	MDF	GLG

SCALE:	MATERIAL:
FULL	CDS

REV.	DESCRIPTION	DATE	BY

UNLESS OTHERWISE SPECIFIED TOLERANCES ARE:

| FRACTIONAL | $\pm$ 1/64 | 2 PLC. DECIMAL $\pm$ .01 |
| 3 PLC. DECIMAL | $\pm$ .005 | 4 PLC. DECIMAL $\pm$ .0005 |

LIMITS ON ANGULAR DIMENSIONS $\pm$ 1/2°

FINISH: BREAK ALL SHARP CORNERS

PART NAME:
SPACER SHIM

PART NUMBER:
A04264415

D-4

6 UNIT
Two-View Drawings

As more complex cylindrical or flat objects are drawn, more than one view is required. Often two views are needed to fully understand the drawing. The drafter must select the two views that will best describe the shape of the object in greatest detail. Usually combinations of the front and right-side or front and top views are used. The selection is often made by eliminating the unnecessary view. An unnecessary view is one that repeats the shape description of another view, Figure 6.1.

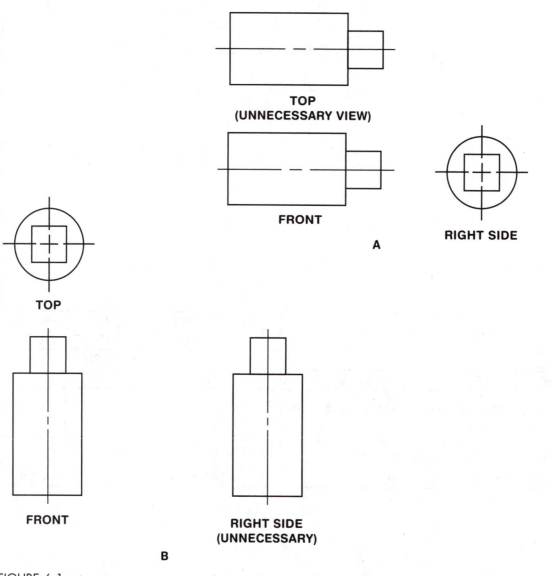

FIGURE 6.1 ■ An unnecessary view repeats the shape description of another view

PROJECTING CYLINDRICAL WORK

Cylindrical pieces that may require more than one view include shafts, collars, studs, and bolts. The view chosen for the front shows the length and shape of the object. The top view or side view describes the object as it might be seen looking at the end. A top or side view of the object shows the circular shape of the part.

The cylindrical piece always has a centerline through its axis. In the circular view, a small cross indicates the center of the object, Figure 6.2. The dimension of the diameter is generally given in the same view as the length of the object, Figure 6.3.

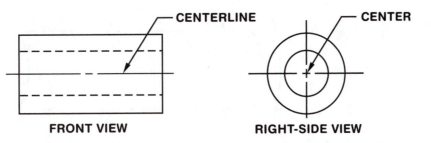

FIGURE 6.2 ■ Cylindrical work is shown in two views

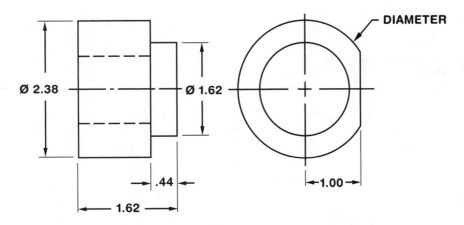

FIGURE 6.3 ■ Dimensioning cylindrical work

SKETCH S-2: STUB SHAFT

1. Using the centerlines provided, sketch front and right-side views of the stub shaft. Allow 1 inch between views.

2. Dimension the completed sketch.

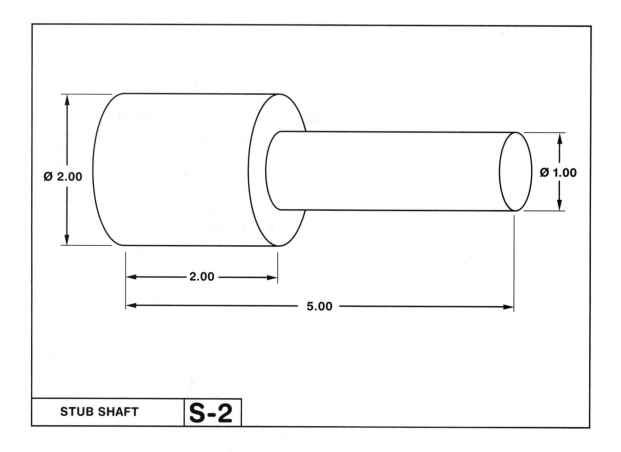

| STUB SHAFT | S-2 |

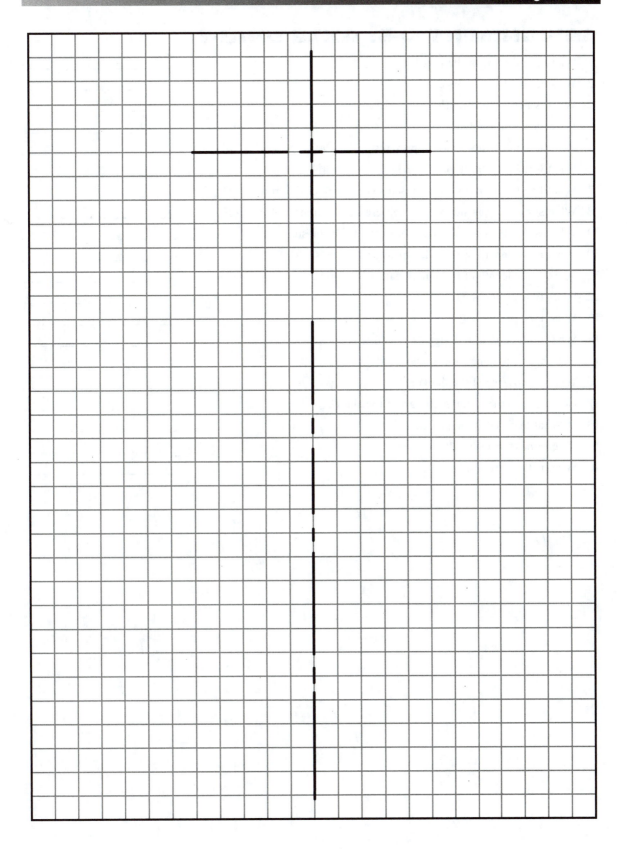

ASSIGNMENT D-5: AXLE SHAFT

1. What is the name of the part? _____

2. What is the date on this drawing? _____

3. What is the overall length? _____

4. What views are shown? _____

5. What line in the front view represents surface Ⓐ? _____

6. What line in the front view represents surface Ⓑ? _____

7. What line in the side view represents surface Ⓒ of the front view? _____

8. What surface in the side view is represented by line Ⓕ of the front view? _____

9. What is the diameter of the cylindrical part of the shaft? _____

10. What is the length of the cylindrical part of the shaft? _____

11. What is the length of the square part of the shaft? _____

12. What type of line is Ⓕ? _____

13. What type of line is Ⓖ? _____

14. What term is used to designate the dimension Ⓙ? _____

15. Determine the dimension shown at Ⓗ. _____

16. What is the scale of the drawing? _____

17. How thick is the rectangular part of the shaft? _____

18. Of what material is the part made? _____

19. Determine dimension Ⓙ. _____

20. What is the part number? _____

Ø 1.375

4.250

7.12

2.12

2.12

				DATE	DWN BY:	CKD BY:	APPR. BY:
				3/14/95	FEC	BDL	JAG
				SCALE:		MATERIAL:	
REV.	DESCRIPTION	DATE	BY	HALF		AL	

STANDARD TOLERANCES
UNLESS OTHERWISE SPECIFIED

FRACTIONAL ± 1/64	2 PLC. DECIMAL ± .01	
3 PLC. DECIMAL ± .005	4 PLC. DECIMAL ± .0005	

LIMITS ON ANGULAR DIMENSIONS ± 1/2°
FINISH: BREAK ALL SHARP CORNERS

PART NAME:
AXLE SHAFT

PART NUMBER:
A05260019

D-5

7 UNIT
Three-View Drawings

One of the main methods of drawing an object is to project it on the sheet in three views. Usually the object is drawn as it is seen by looking directly at the front, top, and right side. In making a drawing, those surfaces of the object that describe its shape are shown in each view.

Figure 7.1 shows how a typical three-view drawing is developed. Figure 7.2 shows how the object would appear on the final drawing.

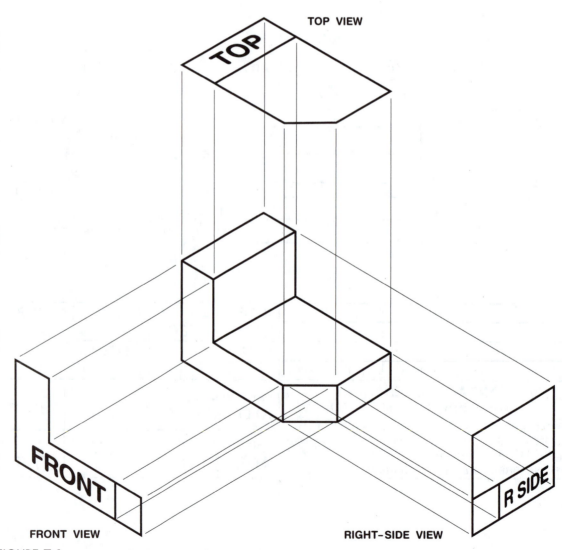

TOP VIEW

TOP

FRONT

FRONT VIEW

R SIDE

RIGHT-SIDE VIEW

FIGURE 7.1 ■ Projecting the three principal views

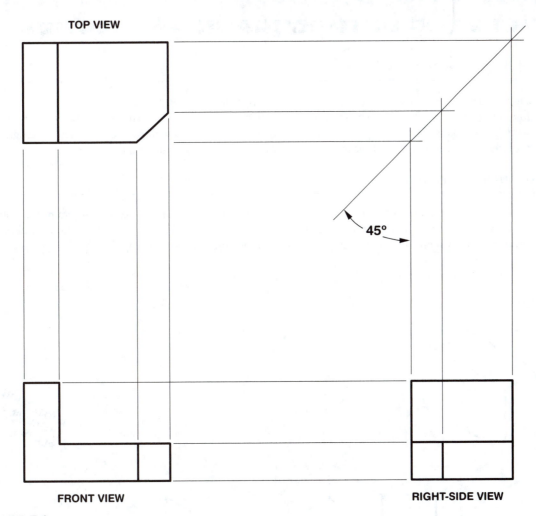

FIGURE 7.2 ■ A typical three-view drawing

FRONT VIEW

The front view of an object is the view that shows the greatest detail of the part. It generally is the view of the object that will provide the most descriptive shape information. It is not necessarily the front of the part as it is used in actual operation.

TOP VIEW

The top view is located directly over and in line with the front view. It is drawn as if the drafter were standing over the object looking straight down.

RIGHT-SIDE VIEW

The right-side view is the most common side view used in a three-view drawing. It is drawn to the right and in line with the front view.

OTHER VIEWS

If the object is even more complex, other views may also be needed. In later units, other views are shown in more detail.

ANGLES OF ORTHOGRAPHIC PROJECTION

Third Angle Projection

Third angle projection is the recognized standard in the United States, Great Britain, and Canada. This system places the projection or viewing plane between the object and the observer, Figure 7.3. The relative position of views in third angle projection is shown in Figure 7.4. The top view is placed directly above the front view. The right-side view is placed directly to the right of the front view. Third angle projection is used in this text.

First Angle Projection

First angle projection, Figure 7.5, is used in most countries other than the United States, Great Britain, and Canada. First angle projection views place the object above the horizontal viewing plane and in front of the vertical projection plane. In first angle projection, views appear as though the observer were looking through the object.

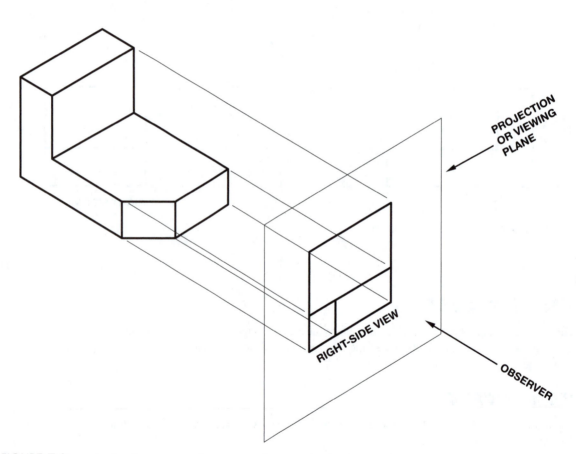

FIGURE 7.3 ■ In third angle projection, the projection (viewing plane) is between the object and the observer

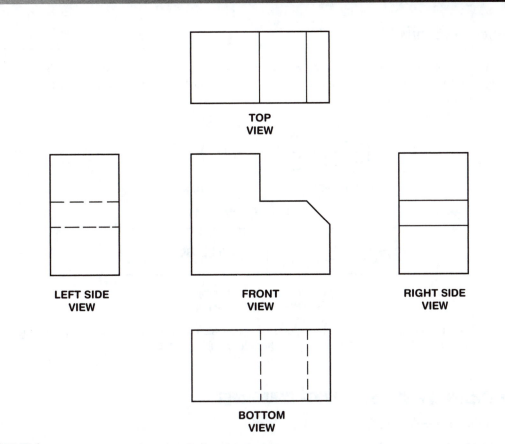

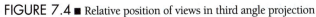

FIGURE 7.4 ■ Relative position of views in third angle projection

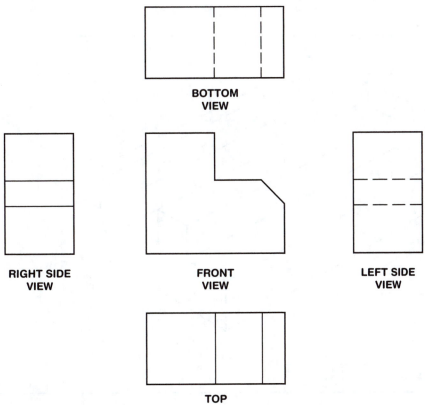

FIGURE 7.5 ■ Relative position of views in first angle projection

ISO Projection Symbols

To indicate the type of projection used on a drawing, a symbol is used. This symbol, which was developed by the International Standards Organization (ISO), is found in or near the title block area of the drawing. Figure 7.6 shows the standard ISO symbol used on drawings.

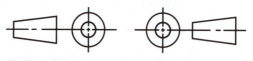

FIRST-ANGLE   **THIRD-ANGLE**

A. ISO projection symbols

TITLE BLOCK

**B. Locating a third angle projection
ISO symbol on drawing paper**

FIGURE 7.6 ■ Standard ISO projection symbols (Reprinted, by permission, from Jensen & Hines, *Interpreting Engineering Drawings, Metric Edition*, Figs. 1.5 and 1.6 by Delmar Publishers Inc.)

SKETCH S-3: SLIDE GUIDE

1. Lay out the front, right-side, and top views.

2. Start the drawing 1/2 inch from the left-hand margin and about 1/4 inch from the bottom. Make the views 3/4 inch apart.

3. Dimension the completed drawing.

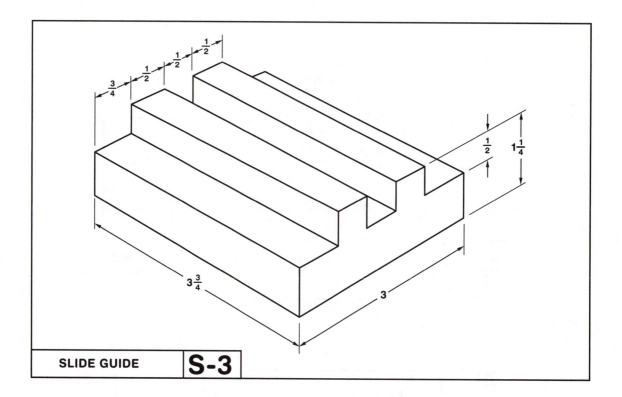

SLIDE GUIDE   **S-3**

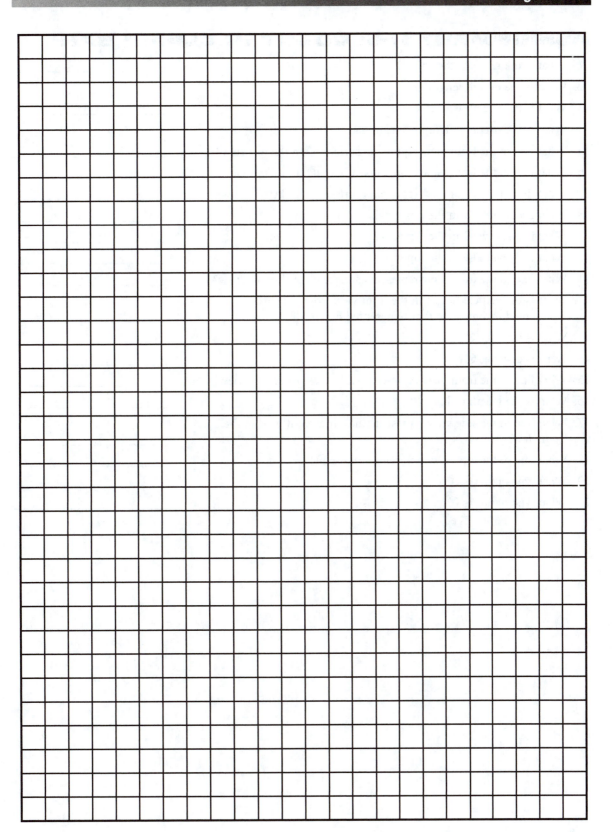

ASSIGNMENT D-6: COUNTER CLAMP BAR

1. What is the name of the object? _____

2. What is the overall length? _____

3. What is the overall width? _____

4. What is the overall height or thickness? _____

5. What line in the front view represents surface Ⓕ in the top view? _____

6. What line in the front view represents surface Ⓔ in the top view? _____

7. What line in the front view represents surface Ⓖ in the top view? _____

8. What is distance Ⓒ in the top view? _____

9. What is distance Ⓓ in the top view? _____

10. What is distance Ⓑ in the top view? _____

11. What line in the front view represents surface Ⓛ of the side view? _____

12. What is the vertical height in the front view from the surface represented by line Ⓟ to that represented by line Ⓠ? _____

13. What is distance Ⓥ? _____

14. What is distance Ⓦ? _____

15. What line in the front view represents surface Ⓜ in the side view? _____

16. What is the length of line Ⓝ? _____

17. What line in the side view represents the surface outlined in the front view? _____

18. What line in the top view represents surface Ⓛ? _____

19. What type of line is Ⓣ? _____

20. What type of line is Ⓨ? _____

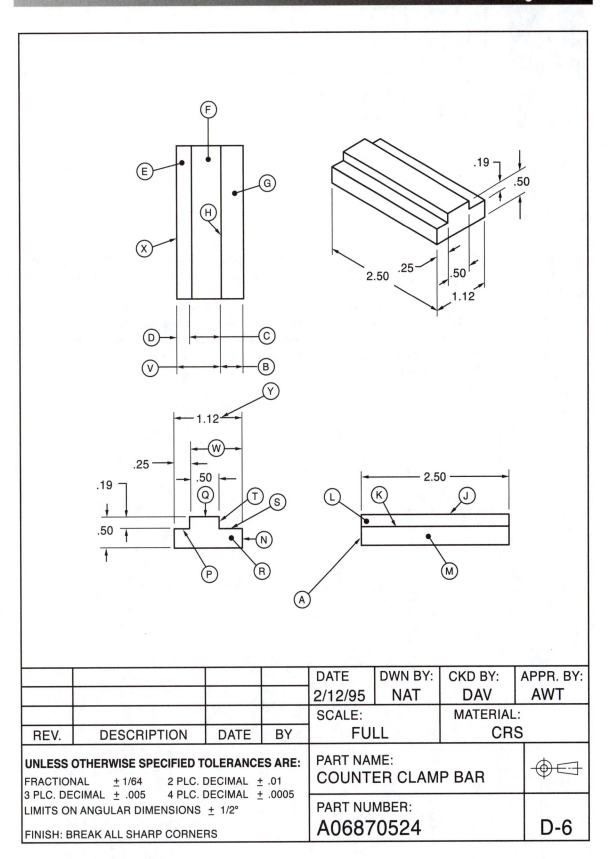

				DATE	DWN BY:	CKD BY:	APPR. BY:
				2/12/95	NAT	DAV	AWT
				SCALE:		MATERIAL:	
REV.	DESCRIPTION	DATE	BY	FULL		CRS	

| UNLESS OTHERWISE SPECIFIED TOLERANCES ARE: | PART NAME:
COUNTER CLAMP BAR | |
| FRACTIONAL ± 1/64   2 PLC. DECIMAL ± .01
3 PLC. DECIMAL ± .005   4 PLC. DECIMAL ± .0005
LIMITS ON ANGULAR DIMENSIONS ± 1/2°

FINISH: BREAK ALL SHARP CORNERS | PART NUMBER:
A06870524 | D-6 |

8 UNIT
Auxiliary Views

An *auxiliary view* is an extra view in which the true size and shape of an inclined surface of the object is represented. An auxiliary view is drawn when it is impossible to show the true shape of inclined surfaces in the regular front, top, or side views.

For example, in Figure 8.1, surface ⊗ is shown in an extra or auxiliary view. In the regular top and side views, the circular surfaces are elliptical. In the auxiliary view, the circular surfaces are round and true to shape. Note that projection lines 1 through 8 are at a 90 degree angle to the surface being projected. *Partial auxiliary views* show only the slanted surfaces that do not appear in true size and shape in the other views. However, *full auxiliary views* are sometimes used to show the entire object.

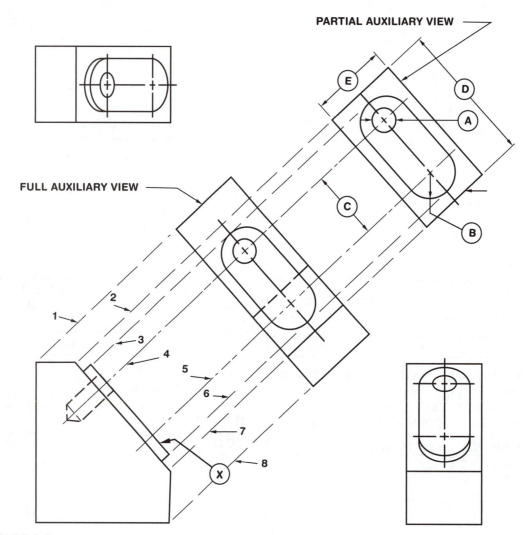

FIGURE 8.1 ■ Auxiliary views

It is accepted practice to show dimensions such as Ⓐ and Ⓑ, Figure 8.1, on the auxiliary view since this is the only place where they occur in their true size and shape. Other dimensions, such as Ⓒ, Ⓓ, and Ⓔ, may also be shown on the auxiliary view if desired.

A primary auxiliary view at Ⓐ in Figure 8.2 shows the true projection and true shape of face Ⓒ. It is projected directly from the regular front view. Another primary auxiliary view is shown at Ⓑ. This view is projected from another side of the regular front view.

To show the true shape and location of the holes, a secondary auxiliary view must be shown. A secondary auxiliary view is projected from a primary auxiliary view and is often required when an object has two or more inclined surfaces.

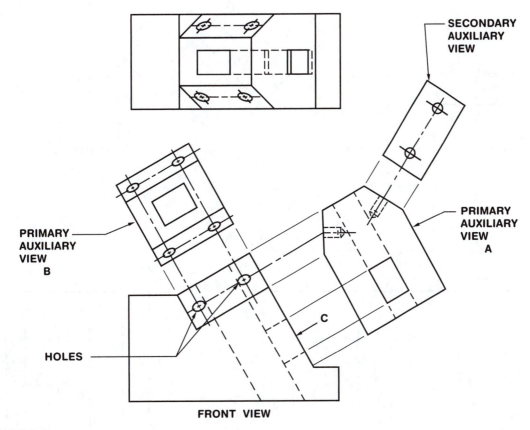

FIGURE 8.2 ■ Primary and secondary auxiliary views

SKETCH S-4: ANGLE BRACKET

Sketch the top auxiliary view for the angle bracket.

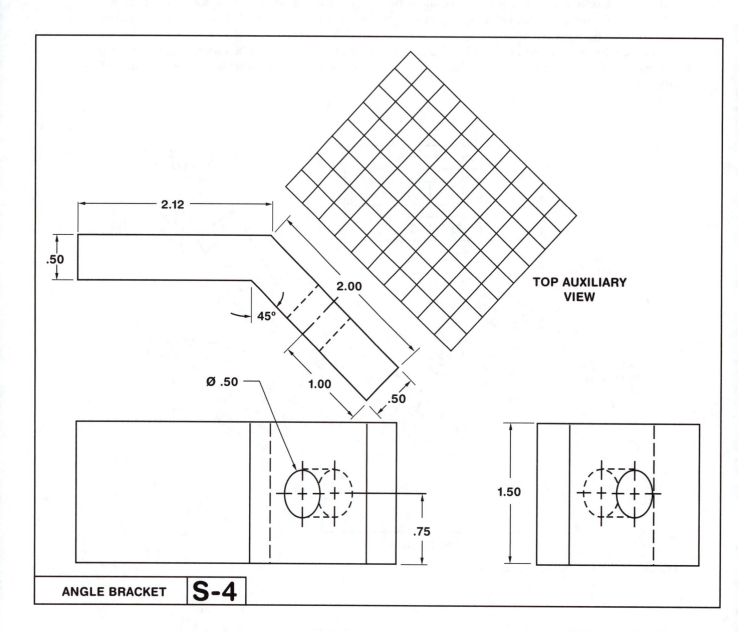

TOP AUXILIARY VIEW

2.12

.50

2.00

45°

Ø .50

1.00

.50

.75

1.50

ANGLE BRACKET   **S-4**

ASSIGNMENT D-7: SLIDE BLOCK

1. What view is shown in View I? _____

2. What view is shown in View IV? _____

3. What view is shown in View III? _____

4. How wide is the slide block from left to right in the front view? _____

5. How high is the part in the front view? _____

6. What is the angle of the inclined surface? _____

7. From what view is the partial auxiliary view projected? _____

8. Is it a primary or secondary auxiliary view? _____

9. What is the width of the inclined surface in the side view? _____

10. What size is the hole? _____

11. What is the length of the inclined surface in the front view? _____

12. What is dimension Ⓐ? _____

13. What is dimension Ⓑ? _____

14. Is the hole in the center of the inclined surface? _____

15. What surface in the top view represents Ⓓ in the side view? _____

16. What surface in the side view represents Ⓖ in the top view? _____

17. Which view shows the true size and shape of surface Ⓗ? _____

18. What surface in the side view represents Ⓔ in the front view? _____

19. What surface in the front view represents Ⓖ in the top view? _____

20. Of what material is the slide block made? _____

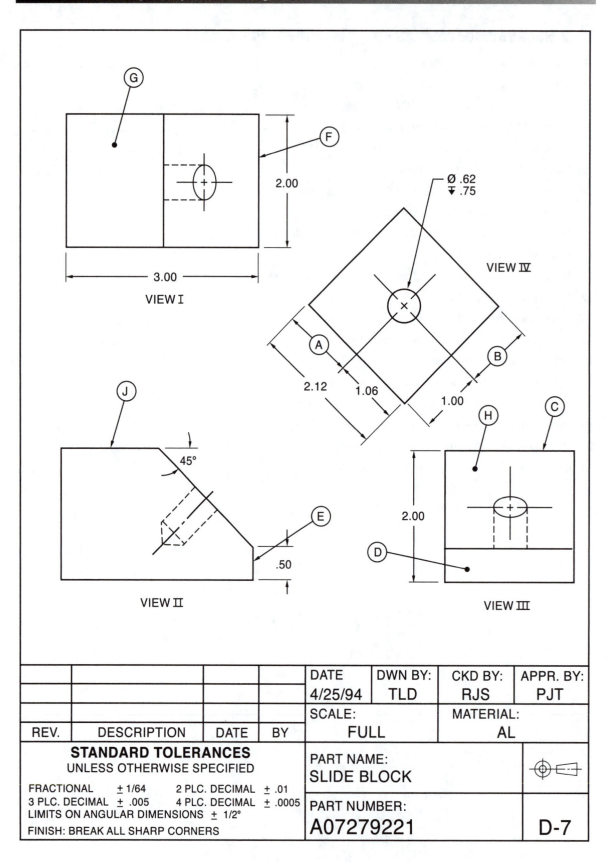

Ⓖ

Ⓕ

2.00

3.00

VIEW I

Ø .62
▼ .75

VIEW Ⅳ

Ⓐ

Ⓑ

2.12   1.06

1.00

Ⓙ

45°

Ⓔ

.50

VIEW II

Ⓗ

Ⓒ

2.00

Ⓓ

VIEW III

				DATE	DWN BY:	CKD BY:	APPR. BY:
				4/25/94	TLD	RJS	PJT
				SCALE:		MATERIAL:	
REV.	DESCRIPTION	DATE	BY	FULL		AL	

STANDARD TOLERANCES
UNLESS OTHERWISE SPECIFIED

FRACTIONAL   ± 1/64   2 PLC. DECIMAL ± .01
3 PLC. DECIMAL ± .005   4 PLC. DECIMAL ± .0005
LIMITS ON ANGULAR DIMENSIONS ± 1/2°
FINISH: BREAK ALL SHARP CORNERS

PART NAME:
SLIDE BLOCK

PART NUMBER:
A07279221

D-7

UNIT 9

Dimensions and Tolerances

An industrial drawing should provide the required information about the size and shape of an object. The print reader must be able to visualize the completed part described on the drawing. In previous units, various views that are used to show the shape of an object have been explained. However, a complete size description is also needed to understand what the machining requirements are.

The size of an object is shown by placing measurements, called *dimensions*, on the drawing. Each dimension has limits of accuracy within which it must fall. These limits are called *tolerances*.

In the following units, the types of dimensions and tolerances used on industrial drawings are discussed.

DIMENSIONS

The size requirements on a drawing may be given in any one or a combination of measuring systems. Dimensions may be fractional, decimal, metric, or angular. Each system will be discussed in detail in later units.

As explained in Unit 3 *Lines and Symbols,* there are special types of lines used in dimensioning. They are called *extension lines, dimension lines,* and *leader lines.* Each has a specific purpose as it is applied to the drawing.

In industrial practice there are a few rules followed in dimensioning a drawing. Understanding these rules is helpful in drawing interpretations and shop sketching. The most common rules are:

- Drawings should supply only those dimensions required to produce their intended objects.

- Dimensions should not be duplicated on a drawing. If a dimension is provided in one view, it should not be given in the other views. Duplicate or double dimensioning is redundant, permits error, and can lead to confusion in print interpretation.

- Dimensions should be placed between views where possible. This helps in identifying points and surface dimensions in adjacent views.

- Dimensions should be spaced from the outside of the object in order of size, Figure 9.1. Smaller dimensions are placed closer to the parts they dimension.

- Notes should be added to dimensions where drawing clarification is required.

- Dimensions should not be placed on the view, if possible.

- Hidden surfaces should not be dimensioned, if possible.

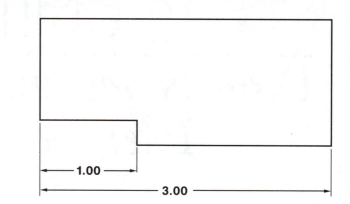

FIGURE 9.1 ■ Dimensions are spaced in order of size

Types of Dimensions

Dimensions placed on drawings are identified as either size or location dimensions. *Size dimensions* are used to indicate lengths, widths, or thicknesses, Figure 9.2. *Location dimensions* are used to show the location of holes, points, or surfaces, Figure 9.3. Both types are often called *construction dimensions*.

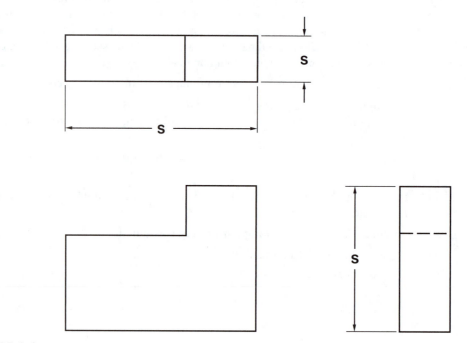

FIGURE 9.2 ■ Size dimensions

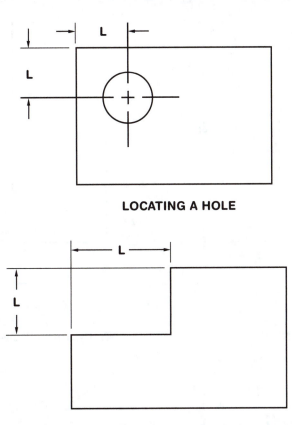

LOCATING A HOLE

LOCATING A SURFACE

FIGURE 9.3 ■ Location dimensions

Methods of Dimensioning

There are two common systems of dimensioning used on industrial drawings. The *aligned method* is read from the bottom and right side of the drawing. To read the dimensions often requires turning the drawing. The aligned method is still used, but it is being replaced with a second system called unidirectional.

The *unidirectional dimensions* are all read from the bottom. Therefore, it is not necessary to turn the print. Figure 9.4 shows the unidirectional and aligned systems of dimensioning.

Note: The most recent ASME standards recommend the use of unidirectional dimensioning. The drawings in this text, therefore, are all shown with unidirectional dimensions.

Reference Dimensions

Reference (calculated) dimensions may be given for information purposes. The old method was to add the letters REF following the dimension to identify it as a reference dimension, Figure 9.4A. New standards require the reference dimension to be placed within parentheses on the drawing, thus eliminating the REF notation, Figure 9.4B.

Dimensions Not to Scale

Objects on original drawings or prints should always be shown true size when possible. There are times, however, when a designer or drafter must insert a dimension or show a surface that is not true size. When a dimension that is not to scale is shown, a straight line is drawn below the dimension.

For example, a dimension shown as 13.875 would not be true length if measured.

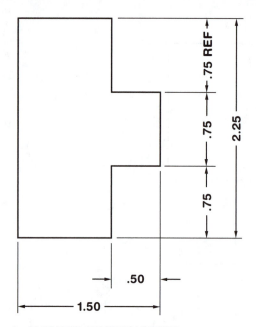

A. ALIGNED DIMENSIONING

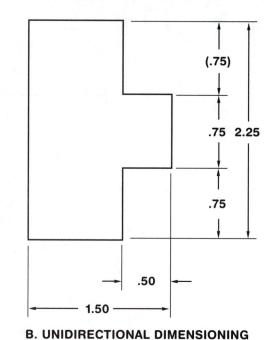

B. UNIDIRECTIONAL DIMENSIONING

FIGURE 9.4 ■ Methods of dimensioning

TOLERANCES

Because it is nearly impossible to make anything to exact size, degrees of accuracy must be specified. When a size is given on a drawing, a tolerance is applied to it. The *tolerance* is a range of sizes within which the actual dimension of a piece must fall. The tolerance specifies how exact the dimension must be.

Just as in dimensioning, the tolerances may be fractional, decimal, or metric. Tolerances may be given in the title block area, Figure 9.5, or on the dimension itself. Tolerances given in the title block apply to all dimensions unless otherwise specified on the drawing.

STANDARD TOLERANCES	
UNLESS OTHERWISE SPECIFIED	
INCH	MILLIMETER
FRACTIONAL ± 1/64 2 PLC. DECIMAL ± .01 3 PLC. DECIMAL ± .005 4 PLC. DECIMAL ± .0005	WHOLE NO. ± 0.5 1 PLC. DECIMAL ± 0.2 2 PLC. DECIMAL ± 0.03 3 PLC. DECIMAL ± 0.013
LIMITS ON ANGULAR DIMENSIONS ± 1/2°	
FINISH: BREAK ALL SHARP CORNERS	

FIGURE 9.5 ■ Dimensional tolerance block

Upper and Lower Limits

All dimensions to which a tolerance is applied have upper and lower limits of size. The *upper limit* is the print dimension with the (+) tolerance added to it. If no (+) tolerance is allowed, the print dimension becomes the upper limit.

The *lower limit* dimension is the print dimension with the (−) tolerance subtracted. If no (−) tolerance is allowed, then the print dimension becomes the lower limit, Figure 9.6.

Methods of Tolerancing

The two systems of tolerancing are known as bilateral and unilateral tolerances. A bilateral tolerance allows for variation in two directions from the print dimension. A tolerance is given as both (+) and (−) minus dimension. For example, 1 3/8 ± 1/64, Figure 9.6A. Bilateral tolerances may not always be an equal amount in each direction. A unilateral tolerance allows for variation in only one direction from the print dimension. The tolerance may be a (+) or a (−) dimension from the print dimension. For example, 1 3/8 + 1/64 or 1 3/8 − 1/64, Figures 9.6B and 9.6C.

FRACTIONAL DIMENSIONS

Perhaps the oldest system used is the fractional system of measurement. This system divides an inch unit into fractional parts of an inch with 1/64 being the smallest fraction used. This is because 1/64 of an inch is the smallest scale dimension that can be read with any degree of accuracy without a magnifying glass.

Fractional dimensioning is used where close tolerances are not required. This is often the case on castings, forgings, standard material sizes, bolts, drilled holes, or machine parts where exact size is unimportant.

FRACTIONAL TOLERANCES

Fractional dimensions usually have a fractional tolerance applied to them. The tolerance may be unilateral or bilateral.

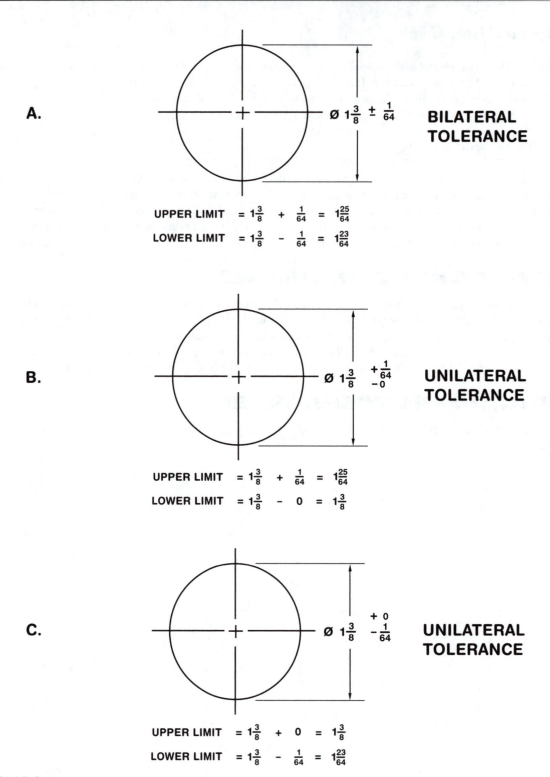

A.

$$\varnothing\ 1\frac{3}{8}\ \pm\ \frac{1}{64}$$

BILATERAL TOLERANCE

UPPER LIMIT $= 1\frac{3}{8}\ +\ \frac{1}{64}\ =\ 1\frac{25}{64}$

LOWER LIMIT $= 1\frac{3}{8}\ -\ \frac{1}{64}\ =\ 1\frac{23}{64}$

B.

$$\varnothing\ 1\frac{3}{8}\ \begin{array}{l}+\frac{1}{64}\\-0\end{array}$$

UNILATERAL TOLERANCE

UPPER LIMIT $= 1\frac{3}{8}\ +\ \frac{1}{64}\ =\ 1\frac{25}{64}$

LOWER LIMIT $= 1\frac{3}{8}\ -\ 0\ =\ 1\frac{3}{8}$

C.

$$\varnothing\ 1\frac{3}{8}\ \begin{array}{l}+\ 0\\-\frac{1}{64}\end{array}$$

UNILATERAL TOLERANCE

UPPER LIMIT $= 1\frac{3}{8}\ +\ 0\ =\ 1\frac{3}{8}$

LOWER LIMIT $= 1\frac{3}{8}\ -\ \frac{1}{64}\ =\ 1\frac{23}{64}$

FIGURE 9.6 ■ Upper and lower limits

ASSIGNMENT D-8: IDLER SHAFT

1. How many hidden diameters are shown in the top view of the idler shaft? _____

2. What is the diameter shown by the hidden line? _____

3. What is the diameter of the smallest visible circle? _____

4. What is the diameter of the largest visible circle? _____

5. What is the length of that portion which is ∅ 1 1/4? _____

6. What is the length of that portion which is ∅ 2 3/4? _____

7. What is the length of that portion which is ∅ 7/8? _____

8. What is the limit of tolerance on fractional dimensions? _____

9. What is the largest size the ∅ 2 3/4 can be turned? _____

10. What is the smallest size the ∅ 2 3/4 can be turned? _____

11. If the length of that portion which is ∅ 1 1/4 is machined to the highest limit, how long will it be? _____

12. If the length of that portion which is ∅ 1 1/4 is machined to 2 1/8 inches, how much over the upper limit of size for this length will it be? _____

13. How much is it over the lower limit, if it is 2 1/8 inches long? _____

14. How long is the shaft from surface A to surface B? _____

15. If the length of the ∅ 2 3/4 measures 2 25/32, will it be over, under, or within the limits of accuracy? _____

16. How much over the lower limit of size will the length of the ∅ 2 3/4 be, if it is 2 25/32? _____

17. What two views of the idler shaft are shown? _____

18. Could the idler shaft have been shown in one view? _____

19. What other two views could have been used? _____

20. What would the overall length of the shaft be if made to the upper limit? _____

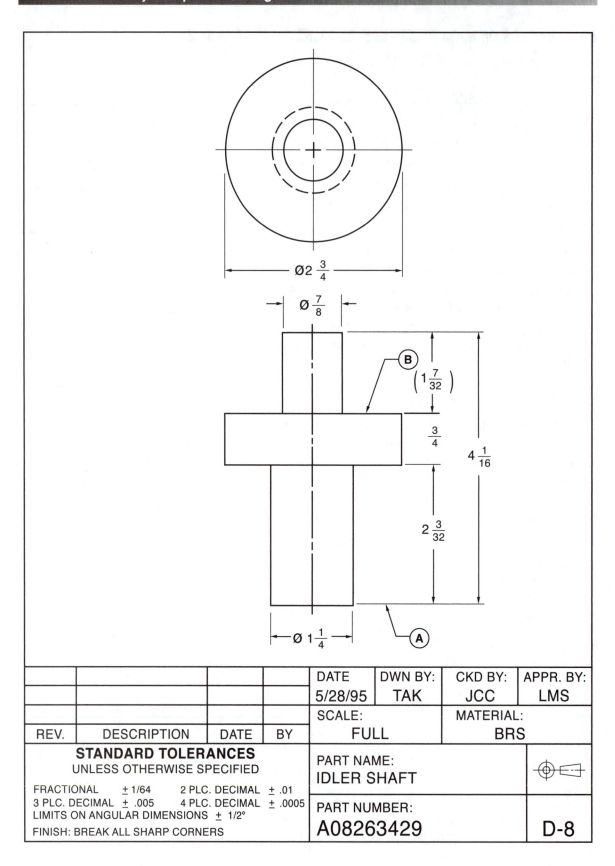

				DATE	DWN BY:	CKD BY:	APPR. BY:
				5/28/95	TAK	JCC	LMS
				SCALE:		MATERIAL:	
REV.	DESCRIPTION	DATE	BY	FULL		BRS	

STANDARD TOLERANCES UNLESS OTHERWISE SPECIFIED	PART NAME: IDLER SHAFT	
FRACTIONAL ± 1/64 2 PLC. DECIMAL ± .01 3 PLC. DECIMAL ± .005 4 PLC. DECIMAL ± .0005 LIMITS ON ANGULAR DIMENSIONS ± 1/2° FINISH: BREAK ALL SHARP CORNERS	PART NUMBER: A08263429	D-8

UNIT 10

Decimal Dimensions and Tolerances

DECIMAL DIMENSIONS

The need for increased accuracy and closer tolerances in machining led to the development of the decimal system. Today the decimal system of measurement has all but replaced the fractional system when high accuracy is desired.

The most common decimal units found on industrial drawings are tenths, hundredths, thousandths, ten-thousandths, and hundred-thousandths. For example:

$$
\begin{aligned}
\text{one-tenth} &= 1/10 &&= 0.10 \\
\text{one-hundredth} &= 1/100 &&= 0.01 \\
\text{one-thousandth} &= 1/1000 &&= 0.001 \\
\text{one-ten-thousandth} &= 1/10,000 &&= 0.0001
\end{aligned}
$$

The unit used depends on the degree of accuracy required for the part. The dimension specified must take into consideration the machining process used. The decimal units on industrial drawings seldom exceed four places for dimensioning. This is due to the fact that machine tools and measuring instruments are usually only accurate to three or four decimal places.

Industrial drawings may be all decimal dimensions or a combination of both decimal and fractional dimensions. The trend, however, has been to dimension totally in decimals to avoid the confusion of using both systems.

Decimal dimensions are preferred because they are easier to work with. They may be added, subtracted, divided, and multiplied with fewer problems in calculation. Decimal numbers may also be directly applied to shop measuring tools, machine tool graduation, modern digital readouts, and computer plots.

DECIMAL TOLERANCES

Just as there are tolerances on fractional dimensions, there are tolerances on decimal dimensions. Decimals in thousandths of an inch are used when greater precision and less tolerance is required to make a part.

Decimal tolerances may be specified in the tolerance block or on a drawing in various ways, Figure 10.1.

POINT-TO-POINT DIMENSIONS

Most linear (in line) dimensions apply on a point-to-point basis. Point-to-point dimensions are applied directly from one feature to another, Figure 10.2. Such dimensions are intended to locate surfaces and features directly between the points indicated. They also locate corresponding points on the indicated surfaces.

For example, a diameter applies to all diameters of a cylindrical surface. It does not merely apply to the diameter at the end where the dimension is shown. A thickness applies to all opposing points on the surfaces.

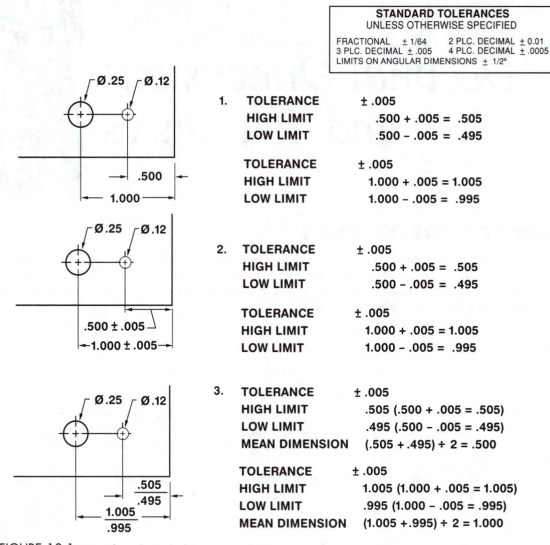

STANDARD TOLERANCES
UNLESS OTHERWISE SPECIFIED
FRACTIONAL ± 1/64   2 PLC. DECIMAL ± 0.01
3 PLC. DECIMAL ± .005   4 PLC. DECIMAL ± .0005
LIMITS ON ANGULAR DIMENSIONS ± 1/2°

1. **TOLERANCE**   ± .005
    HIGH LIMIT   .500 + .005 = .505
    LOW LIMIT   .500 – .005 = .495

    TOLERANCE   ± .005
    HIGH LIMIT   1.000 + .005 = 1.005
    LOW LIMIT   1.000 – .005 = .995

2. **TOLERANCE**   ± .005
    HIGH LIMIT   .500 + .005 = .505
    LOW LIMIT   .500 – .005 = .495

    TOLERANCE   ± .005
    HIGH LIMIT   1.000 + .005 = 1.005
    LOW LIMIT   1.000 – .005 = .995

3. **TOLERANCE**   ± .005
    HIGH LIMIT   .505 (.500 + .005 = .505)
    LOW LIMIT   .495 (.500 – .005 = .495)
    MEAN DIMENSION   (.505 + .495) ÷ 2 = .500

    TOLERANCE   ± .005
    HIGH LIMIT   1.005 (1.000 + .005 = 1.005)
    LOW LIMIT   .995 (1.000 – .005 = .995)
    MEAN DIMENSION   (1.005 + .995) ÷ 2 = 1.000

FIGURE 10.1 ■ Specifying decimal tolerances on a drawing

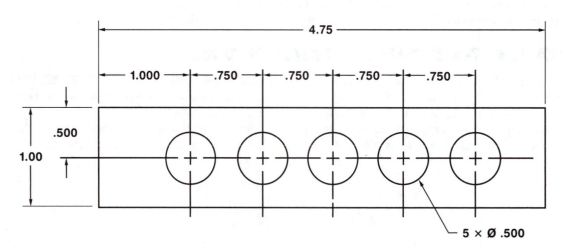

FIGURE 10.2 ■ Point-to-point dimensioning

RECTANGULAR COORDINATE DIMENSIONING

Rectangular coordinate dimensioning, often called datum or baseline dimensioning, is a system where dimensions are given from one or more common data points. Linear dimensions are typically specified from two or three perpendicular planes, Figure 10.3.

Rectangular coordinate dimensioning is often used when accurate part layout is required. Having common data points helps overcome errors that may accumulate in the build-up of tolerances in between point-to-point dimensions.

The datum used in Figure 10.4 is a centerline. The tolerances from the datum must be held to one-half the tolerance acceptable between surface features.

RECTANGULAR COORDINATE DIMENSIONING WITHOUT DIMENSION LINES

Dimensions may also be given without the use of dimension lines. The datum is often shown as a zero line and the dimensions are placed on the extension lines, Figure 10.5.

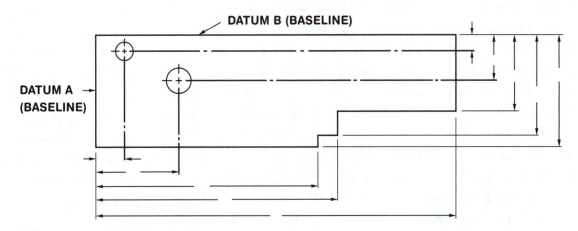

FIGURE 10.3 ■ Rectangular coordinate dimensioning

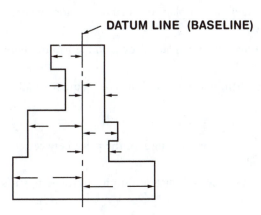

FIGURE 10.4 ■ The centerline is the datum line or baseline in this illustration

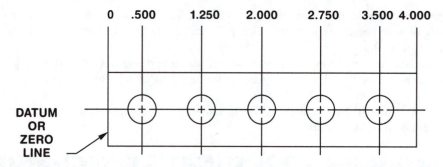

0   .500   1.250   2.000   2.750   3.500 4.000

**DATUM
OR
ZERO
LINE**

FIGURE 10.5 ■ Rectangular coordinate dimensioning without dimension lines

ASSIGNMENT D-9: LOWER DRUM SHAFT

1. What is the overall length of the lower drum shaft? _____

2. What surface is used as the baseline or datum? _____

3. How many diameters are there? _____

4. What is the distance from surface Ⓒ to surface Ⓓ? _____

5. Of what material is the part made? _____

6. What is the minimum size allowed on the 2.875 diameter? _____

7. What is the maximum size allowed on the 1.250 diameter? _____

8. What is the mean length of the $\varnothing\ 1.250 \begin{smallmatrix}+.002\\-.000\end{smallmatrix}$? _____

9. What is the length of the $\frac{1.000}{.998}$ diameter? _____

10. What is the length or thickness of the $\varnothing$ 2.875 collar? _____

11. What is the length from surface Ⓐ to surface Ⓒ? _____

12. What is the length from surface Ⓔ to surface Ⓒ? _____

13. Determine the distance between surface Ⓑ and surface Ⓒ. _____

14. How many hidden lines would be drawn in a top view of the part? _____

15. What tolerance is permitted on two-place decimal dimensions where tolerance is not specified? _____

16. What tolerance is permitted on three-place decimal dimensions where tolerance is not specified? _____

17. How many diameters are being held within limits of accuracy smaller than ± .005? _____

18. What is the lower limit of the 2.375 length? _____

19. What is the upper limit of size for the length of the piece between surface Ⓐ and surface Ⓒ? _____

20. What is the lower limit of size for the distance from surface Ⓐ to surface Ⓑ? _____

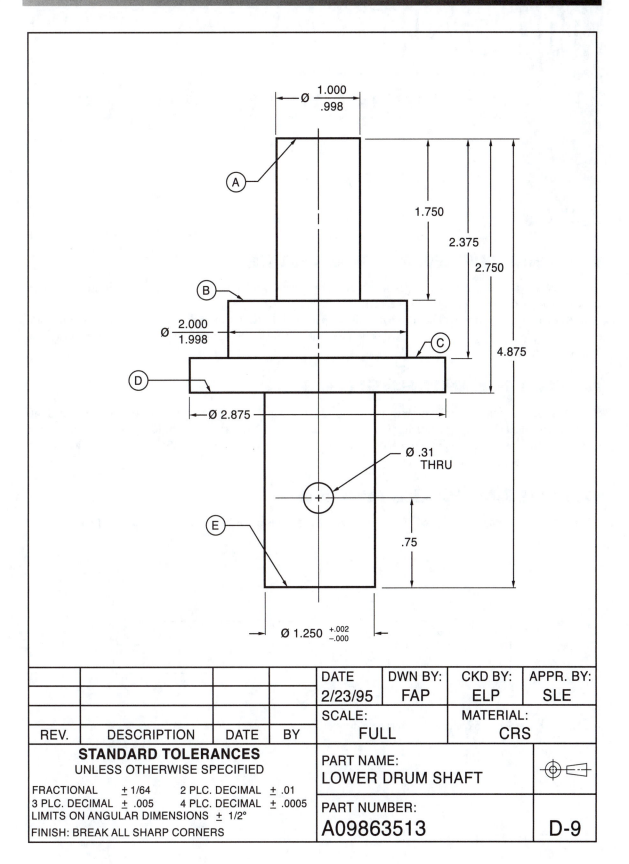

Ø 1.000 / .998

A

1.750

2.375

2.750

4.875

B

Ø 2.000 / 1.998

C

D

Ø 2.875

Ø .31
THRU

E

.75

Ø 1.250 +.002 / −.000

				DATE	DWN BY:	CKD BY:	APPR. BY:
				2/23/95	FAP	ELP	SLE
				SCALE:		MATERIAL:	
REV.	DESCRIPTION	DATE	BY	FULL		CRS	

STANDARD TOLERANCES UNLESS OTHERWISE SPECIFIED	PART NAME: LOWER DRUM SHAFT	
FRACTIONAL ± 1/64 2 PLC. DECIMAL ± .01 3 PLC. DECIMAL ± .005 4 PLC. DECIMAL ± .0005 LIMITS ON ANGULAR DIMENSIONS ± 1/2° FINISH: BREAK ALL SHARP CORNERS	PART NUMBER: A09863513	D-9

Dimensioning Angles

MEASUREMENT OF ANGLES

Some objects do not have all their straight lines drawn horizontally and vertically. The design of the part may require some lines to be drawn at an angle, Figure 11.1.

The amount by which these lines diverge or draw apart is indicated by an *angle dimension*. The unit of measure of such an angle is the *degree* and is denoted by the symbol °. There are 360 degrees in a complete circle. On a drawing, 360 degrees may be written as 360°.

ANGULAR DIMENSIONS

Sizes of angles are dimensioned in degrees. Each degree is 1/360 of a circle. The degree may be further divided into smaller units called *minutes* ('). There are 60 minutes in each degree. The minute may be further divided into smaller units called *seconds* (''). There are 60 seconds in each minute. For example: 10° 15' 35'' would be a typical dimension given in degrees, minutes, and seconds.

ANGULAR TOLERANCES

Angular tolerances may be expressed either on the angular dimension or in a note on the drawing, Figure 11.2.

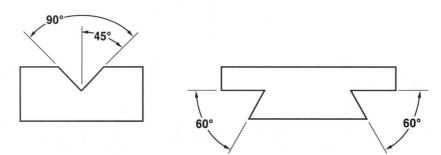

FIGURE 11.1 ■ Some objects may have angular dimensions

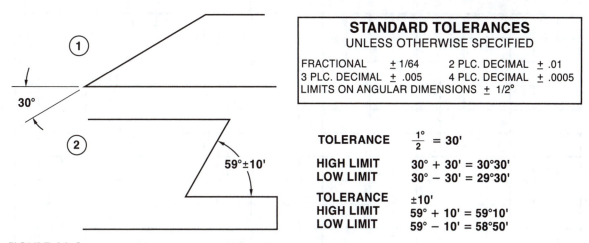

STANDARD TOLERANCES
UNLESS OTHERWISE SPECIFIED

FRACTIONAL	± 1/64	2 PLC. DECIMAL ± .01
3 PLC. DECIMAL	± .005	4 PLC. DECIMAL ± .0005

LIMITS ON ANGULAR DIMENSIONS ± 1/2°

TOLERANCE	$\frac{1°}{2} = 30'$
HIGH LIMIT	30° + 30' = 30°30'
LOW LIMIT	30° − 30' = 29°30'
TOLERANCE	±10'
HIGH LIMIT	59° + 10' = 59°10'
LOW LIMIT	59° − 10' = 58°50'

FIGURE 11.2 ■ ■ Angular tolerances as specified on a drawing

IMPLIED 90 DEGREE ANGLES

Surface features and centerlines intersecting at right angles are not specified with an angular dimension of 90° on the drawing. It is generally understood that lines and surfaces shown to be at right angles will be 90° unless otherwise specified. The standard tolerance for implied 90° angles is the same as the standard tolerance specified for other angular features on the drawing.

SKETCH S-5: REST BRACKET

1. Lay out front, right-side, and top views.

2. Start the sketch 1/2 inch from the left-hand margin and about 1/2 inch from the bottom. Make the views 1 inch apart.

3. Dimension the completed drawing.

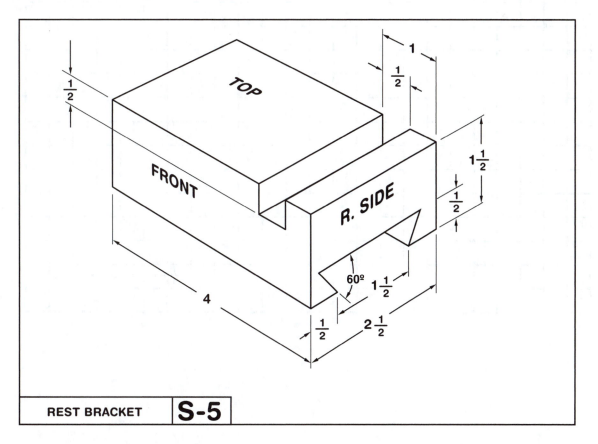

REST BRACKET   **S-5**

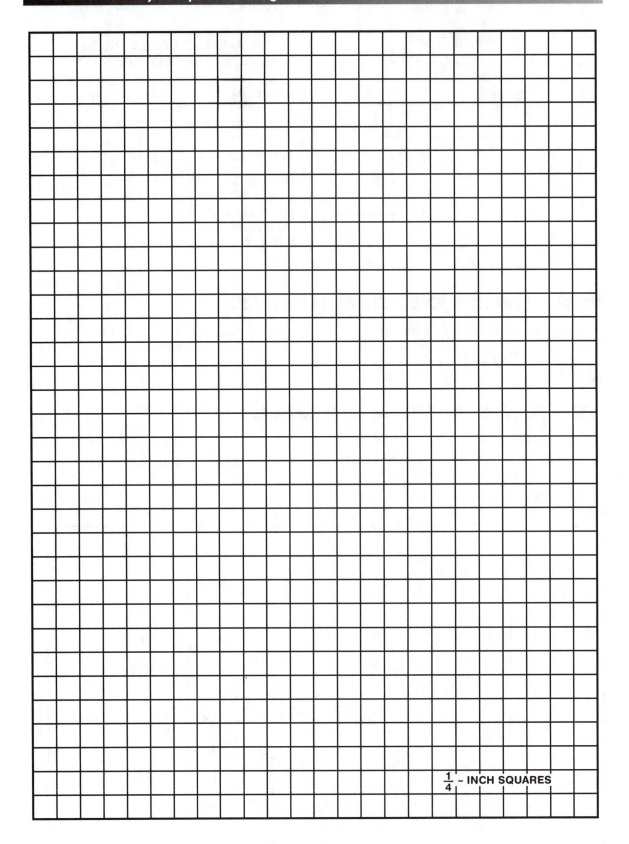

$\frac{1}{4}$ – INCH SQUARES

ASSIGNMENT D-10: FEED HOPPER

1. What is the overall length of the feed hopper? _____

2. What is the overall height or thickness? _____

3. What surface in the top view represents surface Ⓒ in the front view? _____

4. What surface in the front view represents surface Ⓝ in the side view? _____

5. What surface in the front view represents surface Ⓡ in the top view? _____

6. What surfaces of the top view represent line Ⓚ of the side view? _____

7. What surface in the top view represents line Ⓜ in the side view? _____

8. What line of the side view does point Ⓧ in the front view represent? _____

9. What surface in the top view does line Ⓖ in the front view represent? _____

10. What surface in the top view does line Ⓔ in the front view represent? _____

11. What surface in the front view represents surface Ⓣ in the top view? _____

12. In the front view, what is the width of the opening at the top of the slot? _____

13. What line in the front view represents surface Ⓢ in the top view? _____

14. What surface in the top view represents line Ⓐ in the front view? _____

15. How far is it from the top of the hopper to the bottom of the slot in the front view? _____

16. What is the distance from the bottom of the slot to the base of the hopper in the front view? _____

17. What is the unit of measurement of angles? _____

18. At what angle from the vertical are the inclined edges of the slot cut? _____

19. If the angle is given a tolerance of ±1/2°, how many minutes would this be? _____

20. What would the upper limit dimension of the 30° angle be using the ±1/2° tolerance? _____

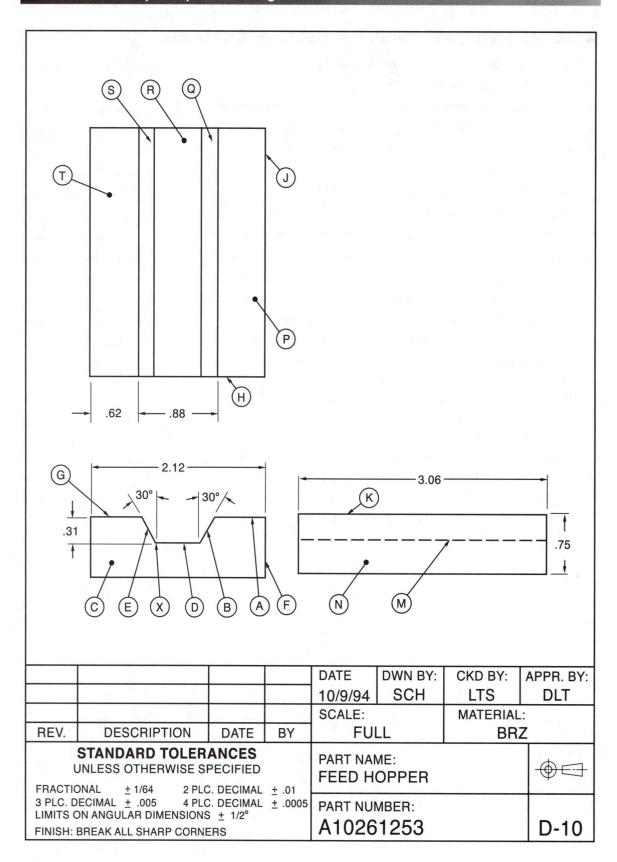

				DATE	DWN BY:	CKD BY:	APPR. BY:
				10/9/94	SCH	LTS	DLT
				SCALE:		MATERIAL:	
REV.	DESCRIPTION	DATE	BY	FULL		BRZ	

STANDARD TOLERANCES UNLESS OTHERWISE SPECIFIED	PART NAME: FEED HOPPER	⊕ ⇥
FRACTIONAL ± 1/64 2 PLC. DECIMAL ± .01 3 PLC. DECIMAL ± .005 4 PLC. DECIMAL ± .0005 LIMITS ON ANGULAR DIMENSIONS ± 1/2° FINISH: BREAK ALL SHARP CORNERS	PART NUMBER: A10261253	D-10

Dimensioning Holes

DIAMETERS OF HOLES

Holes in objects may be dimensioned in several ways. The method used often depends on the size of the hole or how it is to be produced. Common dimensioning practice is to dimension holes on the view in which they appear as circles. Hole diameters should not be dimensioned on the view in which they appear as hidden lines.

Small hole sizes are shown with a leader line. The leader touches the outside diameter of the hole and points to the center. The hole diameter is given at the end of the leader outside the view of the object.

DEPTH OF HOLES

Holes that go through a part may not require any additional information other than a diameter or repetitive feature call-out, Figure 12.1. When it is not clear that a hole goes through the part, however, the letters THRU may follow the hole diameter dimension specified.

Holes that do not go through a part are commonly referred to as blind holes and must have a note or symbol specifying the depth. The depth of a hole is defined as the length of the full diameter of the hole. It is not the depth of the hole from the outer surface to the point of the drill, for example. In the past a note was added to the diameter specification indicating the depth of the hole. Current ASME dimensioning standards require the use of a depth symbol, Figure 12.2.

Large holes may be dimensioned within the diameter of the circle on the view, Figure 12.3.

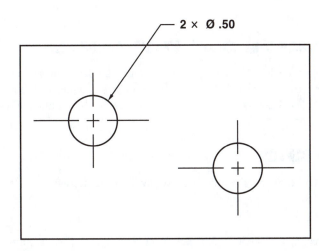
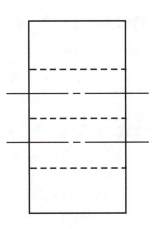

FIGURE 12.1 ■ Hole diameter call-out

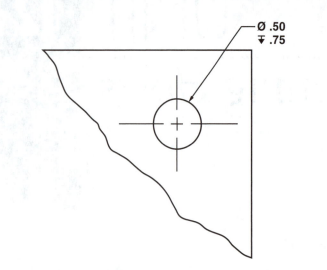
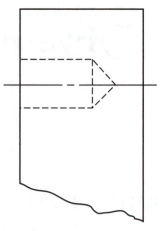

FIGURE 12.2 ■ Hole diameter and depth call-out

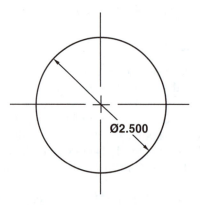

FIGURE 12.3 ■ Hole diameter specified within large hole

LOCATION OF CIRCULAR HOLE PATTERNS

Holes are often spaced in a circular pattern on an object. Each hole location in the pattern shares a common centerline. The circle formed by the centerline is often referred to as the *bolt circle*. The bolt circle is dimensioned by giving the diameter of the circle. Holes located on the hole circle may be equally spaced from each other or unequally spaced.

HOLES EQUALLY SPACED

A circle contains 360 degrees. Holes may be located around the circumference by dividing the number of holes required into the number of degrees. For example:

4 holes equally spaced on a circle = 360° ÷ 4 = 90°

The locations of the holes on the circle are 90 degrees apart. The hole diameter, spacing in degrees, and the number of holes of that size required are given by notes and symbols at the end of a leader line, Figure 12.4. Additional leaders and notes are required for each different size hole required.

HOLES UNEQUALLY SPACED

Holes unequally spaced on a circle are usually located by means of angular dimensions. The angular dimensions use a common centerline for reference to aid in proper hole location, Figure 12.5. The diameter of the bolt circle is also provided with a diagonal dimension line.

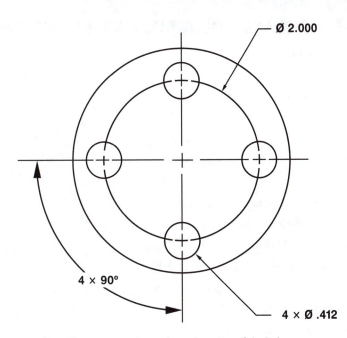

FIGURE 12.4 ■ The notes indicate the size, quantity, and *equal* spacing of the holes

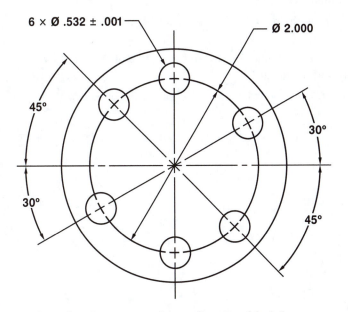

FIGURE 12.5 ■ The notes indicate the size, quantity, and *unequal* spacing of the holes

ASSIGNMENT D-11: FLANGED BUSHING

1. How many ∅ .25 holes are in the bushing? _____

2. How far apart are they? _____

3. On what diameter circle are they located? _____

4. What type of line is Ⓐ? _____

5. What type of line is Ⓔ? _____

6. What lines in the front view represent diameter Ⓓ? _____

7. What kind of lines are used to represent diameter Ⓓ in the front view? _____

8. What line in the front view represents surface Ⓕ? _____

9. What is the diameter of the hole through the center of the bushing? _____

10. What is the minimum allowable diameter for the 1.125 hole? _____

11. What is the maximum allowable diameter for the 1.125 hole? _____

12. What is the tolerance allowed on "unspecified" 3-place decimal dimensions? _____

13. What is the outside diameter of the flange? _____

14. What is the thickness, in fractional dimensions, of the flange? _____

15. What is the length of the body of the bushing? _____

16. What is the longest length the body can be made? _____

17. What is the outside diameter of the body? _____

18. What is the distance from the outside of the body to the outside of the flange? _____

19. Is the centerline for the drilled holes located in the middle of the shoulder formed by the body and the flange? _____

20. What is the distance from the outside of a ∅ .25 hole to the outside of the flange? _____

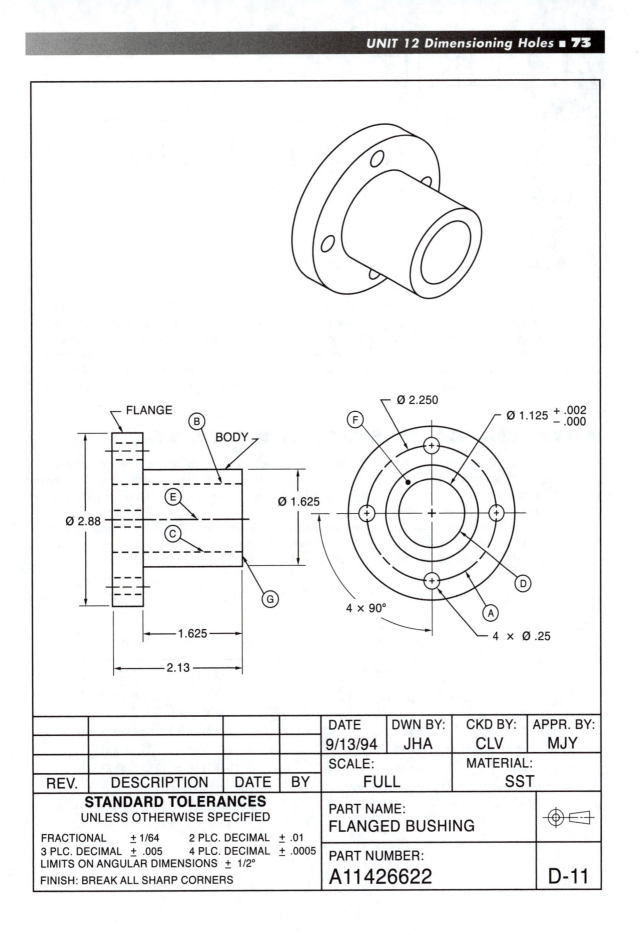

FLANGE

B

BODY

Ø 2.88

E

C

G

1.625

2.13

Ø 1.625

Ø 2.250

F

Ø 1.125 +.002 −.000

4 × 90°

4 × Ø .25

A

D

				DATE	DWN BY:	CKD BY:	APPR. BY:
				9/13/94	JHA	CLV	MJY
				SCALE:		MATERIAL:	
REV.	DESCRIPTION	DATE	BY	FULL		SST	

STANDARD TOLERANCES
UNLESS OTHERWISE SPECIFIED

FRACTIONAL ± 1/64   2 PLC. DECIMAL ± .01
3 PLC. DECIMAL ± .005   4 PLC. DECIMAL ± .0005
LIMITS ON ANGULAR DIMENSIONS ± 1/2°
FINISH: BREAK ALL SHARP CORNERS

PART NAME:
FLANGED BUSHING

PART NUMBER:
A11426622

D-11

13 UNIT
Metric Dimensions

The metric system of measurement is certainly not new to the world. It was first established in France in the late 1700s and has since become the standard of measurement in most countries. However, the traditional system of measurement most familiar in the United States and United Kingdom has been the English system of measurement.

Increased international trade and worldwide use of metrics has caused a reevaluation of the English system of measurement. To be competitive in foreign markets and assure interchangeability of parts, the use of metrics is required. In 1975 the U.S. Metric Conversion Act was signed into law. Since that time, industry has slowly started to change from the English system to the metric system.

INTERNATIONAL SYSTEM OF UNITS (SI)

In 1954, the Conférence Générale des Poids et Mesures (CGPM), which is responsible for all international metric decisions, adopted the Système International d'Unités. The abbreviation for this system is simply SI. The SI metric system establishes the meter as the basic unit of length. Additional length measures are formed by multiplying or dividing the meter by powers of 10, Figure 13.1.

The standard inch measurements on metric drawings are replaced by millimeter dimensions, Figure 13.2. A millimeter is 1/1000 of a meter. Angular dimensions and tolerances remain unchanged in the metric system because degrees, minutes, and seconds are common to both systems of measurement.

1 meter	$= \dfrac{1}{1000}$ kilometers
1 meter	$=$ 10 decimeters
1 meter	$=$ 100 centimeters
1 meter	$=$ 1000 millimeters

FIGURE 13.1 ■ Multiples of the meter

IN		MM
.0001	=	.00254
.001	=	.02540
.010	=	.25400
.100	=	2.54000
1.000	=	25.40000
10.000	=	254.00000

FIGURE 13.2 ■ Inches to millimeters

DIMENSIONING METRIC DRAWINGS

Many industrial drawings contain both English and metric measurements. This practice is called *dual dimensioning*. Dual dimensions provide a reference when converting from one system to the other is required. The most common method of dual dimensioning is to dimension the object using one system of measure while providing conversion information in chart form in a separate area on the drawing, Figure 13.3.

More recent metric drawings omit the dual dimensions and are dimensioned using only metric units. Drawings that are totally metric frequently have a note applied specifying that it is dimensioned in metric units, Figure 13.4.

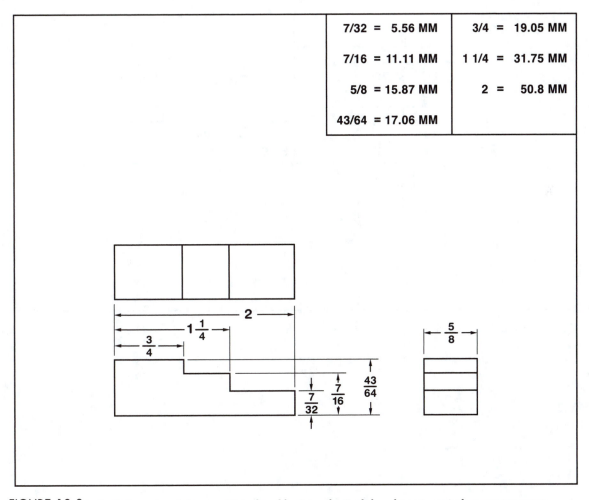

FIGURE 13.3 ■ Dual-dimensioned drawing. Note the table giving the English and metric equivalent measurements.

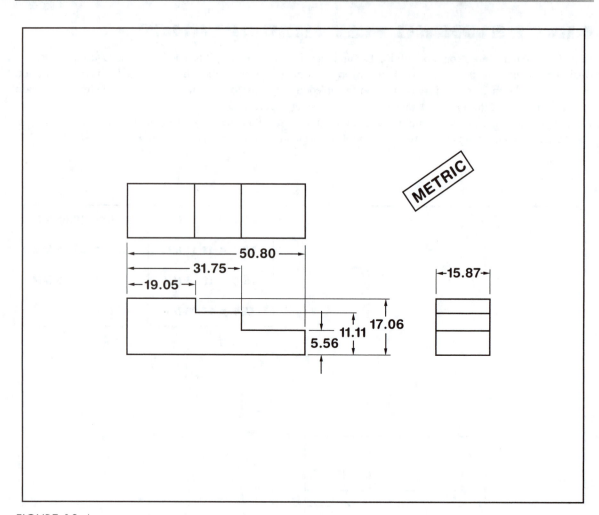

FIGURE 13.4 ■ Drawing dimensioned in metric units of measure only

ASSIGNMENT D-12: LOCATOR

1. What is the overall length of the locator in inches? _____

2. What is the overall length of the locator in millimeters? _____

3. How many 12.7-mm holes are there in the locator? _____

4. What is the overall height of the locator in millimeters? _____

5. What are the dimensions of the chamfer on the corner? _____

6. How thick is the locator in inches? _____

7. How thick is the locator in millimeters? _____

8. What does surface Ⓐ represent? _____

9. What kind of line is shown at Ⓑ? _____

10. What kind of line is shown at Ⓒ? _____

11. What line in the front view represents surface Ⓓ in the top view? _____

12. What line in the top view represents surface Ⓘ in the side view? _____

13. What line in the side view represents surface Ⓓ in the top view? _____

14. What surface in the side view represents line Ⓕ in the front view? _____

15. How many millimeters are there in three inches? _____

16. What is the distance from line Ⓚ to line Ⓒ in millimeters? _____

17. What is the name of the system of measurement that uses inch units? _____

18. What is the official name given to the metric system of measurement? _____

MILLIMETERS	INCHES
9.5	.37
12.7	.50
19.1	.75
25.4	1.00
50.8	2.00
101.6	4.00

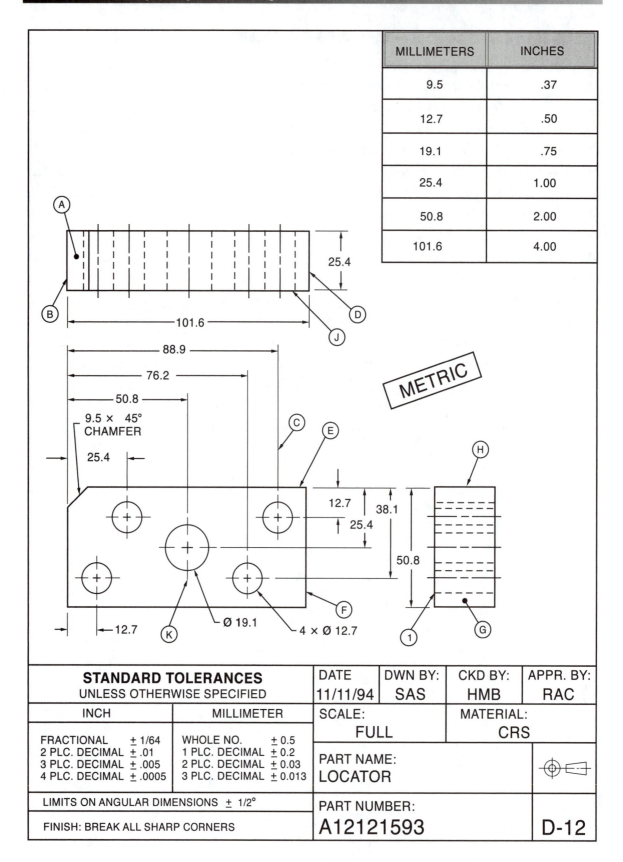

METRIC

9.5 × 45° CHAMFER

Ø 19.1

4 × Ø 12.7

STANDARD TOLERANCES		DATE	DWN BY:	CKD BY:	APPR. BY:
UNLESS OTHERWISE SPECIFIED		11/11/94	SAS	HMB	RAC
INCH	MILLIMETER	SCALE: FULL		MATERIAL: CRS	
FRACTIONAL ± 1/64 2 PLC. DECIMAL ± .01 3 PLC. DECIMAL ± .005 4 PLC. DECIMAL ± .0005	WHOLE NO. ± 0.5 1 PLC. DECIMAL ± 0.2 2 PLC. DECIMAL ± 0.03 3 PLC. DECIMAL ± 0.013	PART NAME: LOCATOR			
LIMITS ON ANGULAR DIMENSIONS ± 1/2°		PART NUMBER:			
FINISH: BREAK ALL SHARP CORNERS		A12121593			D-12

UNIT 14

Full Sections

The details of the interior of an object may be shown more clearly if the object is drawn as though a part of it were cut away, exposing the inside surfaces. When showing an object in section, all surfaces that were hidden are drawn as visible surface lines. The surfaces that have been cut through are indicated by a series of slant lines, known as *section lining*. The line that indicates the plane cutting through the object is the *cutting plane line*, Figure 14.1.

After being cut, the portion of the object to the right of the cutting plane line in Figure 14.1 is considered to be removed. The portion to the left of the cutting plane line is viewed in the direction of the arrows as shown in section A–A.

MATERIALS IN SECTION

In drawing sections of various machine parts, section lines indicate the different materials of which the parts are made, Figure 14.2. Each material is represented by a different pattern of lines. On most drawings, however, sections are shown using the pattern for cast iron. The kind of material is then indicated in the specifications.

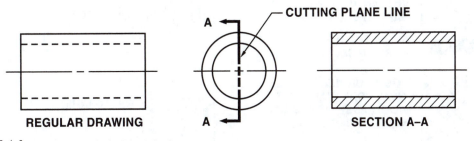

REGULAR DRAWING   SECTION A–A

FIGURE 14.1 ■ Sectioning of a hollow cylinder

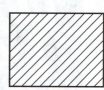 CAST IRON AND GENERAL PURPOSE USE FOR ALL MATERIALS

STEEL

BRONZE, BRASS, COPPER, AND COMPOSITIONS

WHITE METAL, ZINC, LEAD, BABBIT, AND ALLOYS

ALUMINUM AND ALUMINUM ALLOYS

ELECTRIC INSULATION, VULCANITE, FIBER, MICA, BAKELITE, ETC.

SHOW SOLID FOR NARROW SECTIONS

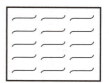 SOUND OR HEAT INSULA-TION, CORK, HAIR-FELT, WOOL, ASBESTOS, MAG-NESIA, PACKING, ETC.

FLEXIBLE MATERIAL FABRIC, FELT RUBBER, ETC.

FIREBRICK AND REFRACTORY MATERIAL

 ELECTRIC WINDINGS, ELECTRO MAGNETS, RESISTANCE, ETC.

 CONCRETE

BRICK OR STONE MASONRY

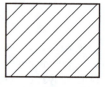 MARBLE, SLATE, GLASS, PORCELAIN, ECT.

EARTH

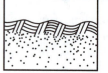 ROCK

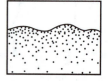 SAND

 WATER AND OTHER LIQUIDS

 ACROSS GRAIN

WITH GRAIN

> WOOD

FIGURE 14.2 ■ Symbols for section lining

SKETCH S-6: COLLARS

1. Copy the sketch of the collars on the grid.
2. Cut through sections A–A and B–B as indicated. Show the front in section.
3. Sketch in section lining.

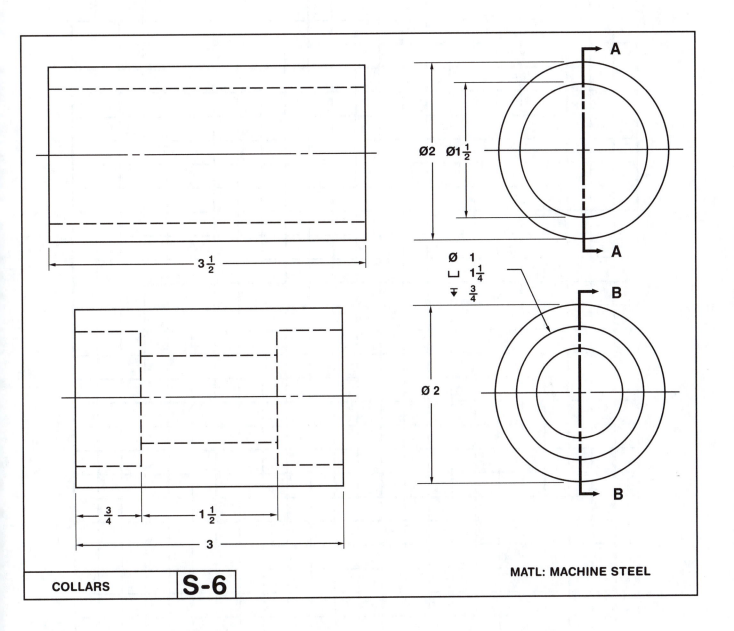

COLLARS   **S-6**

MATL: MACHINE STEEL

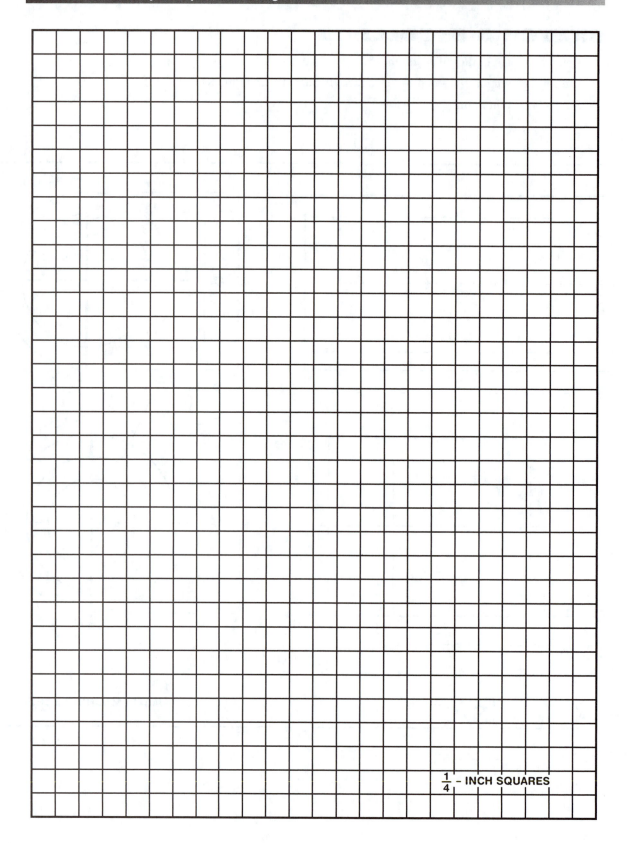

$\frac{1}{4}$ – INCH SQUARES

ASSIGNMENT D-13: TOOL POST BLOCK

1. What heat treatment does the tool post block receive? _____

2. What is the shape of the piece? _____

3. What is the length and the width? _____

4. How thick is it? _____

5. What type of section is shown at A–A? _____

6. What type of line is drawn horizontally through the center of the block in the top view? _____

7. What surface in the top view does line Ⓓ represent? _____

8. What surface in the top view does line Ⓒ represent? _____

9. What is the diameter of the smaller hole in the block? _____

10. How deep is the larger diameter hole? _____

11. How thick is the material between surface Ⓓ and surface Ⓕ? _____

12. What is the upper limit of size of the ∅ 1.343 dimension? _____

13. What is the lower limit of size of the ∅ 1.343 dimension? _____

14. If the tool post that passes through this block is ∅ 1.250, what is the clearance between the sides of the tool post and the smaller hole? _____

15. What finish is specified? _____

16. What fractional tolerances are specified for the tool post block? _____

17. What angular tolerances are specified for the tool post block? _____

18. What is the upper limit of the ∅ 2.000 hole? _____

19. What material does the section lining indicate? _____

20. What radius is required in the bottom of the ∅ 2.000 hole? _____

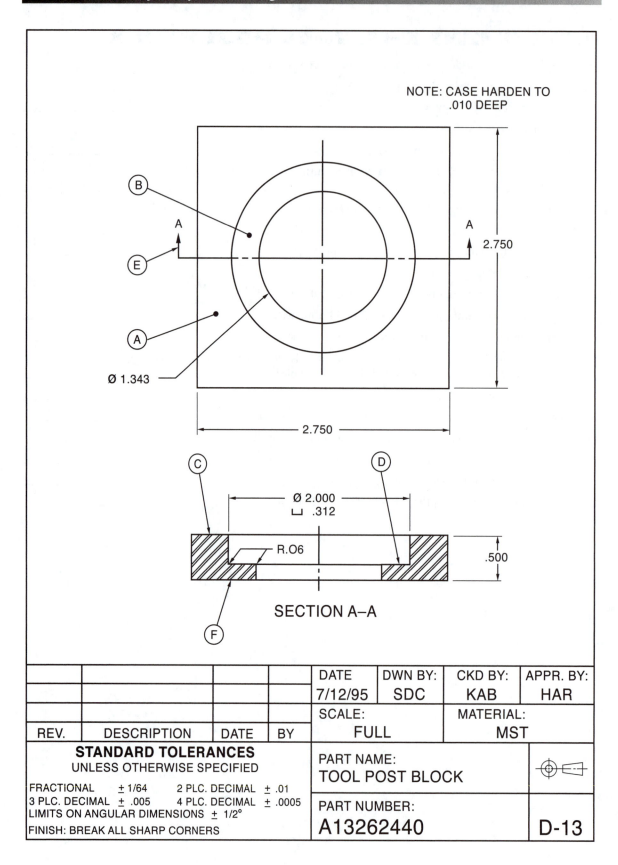

NOTE: CASE HARDEN TO
.010 DEEP

B

A

E

A

Ø 1.343

2.750

2.750

C   D

Ø 2.000
⌴ .312

R.O6

.500

F

SECTION A–A

				DATE	DWN BY:	CKD BY:	APPR. BY:
				7/12/95	SDC	KAB	HAR
				SCALE:		MATERIAL:	
REV.	DESCRIPTION	DATE	BY	FULL		MST	

STANDARD TOLERANCES
UNLESS OTHERWISE SPECIFIED

FRACTIONAL ± 1/64   2 PLC. DECIMAL ± .01
3 PLC. DECIMAL ± .005   4 PLC. DECIMAL ± .0005
LIMITS ON ANGULAR DIMENSIONS ± 1/2°
FINISH: BREAK ALL SHARP CORNERS

PART NAME:
TOOL POST BLOCK

PART NUMBER:
A13262440   D-13

A *half section view* is one in which one half of the view is drawn in section and the other half as a usual exterior view, Figure 15.1. The cut is imagined to extend halfway across, stopping at the axis or centerline. This has the advantage of showing both the exterior and interior on one view. Hidden edges are usually not shown, but they may be shown, if necessary, for dimensioning purposes.

The cutting plane line and letters or arrows designating the direction to view the section are generally eliminated in practice on simple symmetrical objects.

CUTTING PLANE LINE

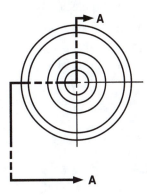

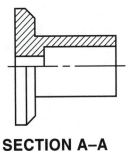

SECTION A–A

FIGURE 15.1 ■ Half section view

ASSIGNMENT D-14: V-BELT PULLEY

1. What is the largest diameter of the idler V-belt pulley? _____

2. What is the overall length of the pulley? _____

3. How wide is the pulley (hub not included)? _____

4. How long is the hub of the pulley? _____

5. What is the diameter of the hub? _____

6. How wide is the V-belt groove on its outside diameter? _____

7. What is the included angle of the V-belt groove? _____

8. How deep is the V-belt groove cut? _____

9. What is the diameter of the hole through the center of the pulley? _____

10. Is the diameter of this hole given in fractional or decimal form? _____

11. What is the diameter of the oil hole? _____

12. How far from the right end is the center of the oil hole located? _____

13. Is the oil hole drilled through one side of the hub to the $\frac{.501}{.502}$ diameter hole? _____

14. How deep is the $\varnothing$ 3/4 recess? _____

15. What material is the V-belt pulley made from? _____

16. What type of section view is shown? _____

17. What scale is used on the drawing of the pulley? _____

18. What is the lower limit of the $\varnothing$ 1 5/8? _____

19. What is the upper limit of the 39° angle? _____

20. If the shaft on which this pulley revolves were turned to $\varnothing$ .499 and the hole in the pulley machined to $\varnothing$ .501, what would be the total clearance between the shaft and the hole? _____

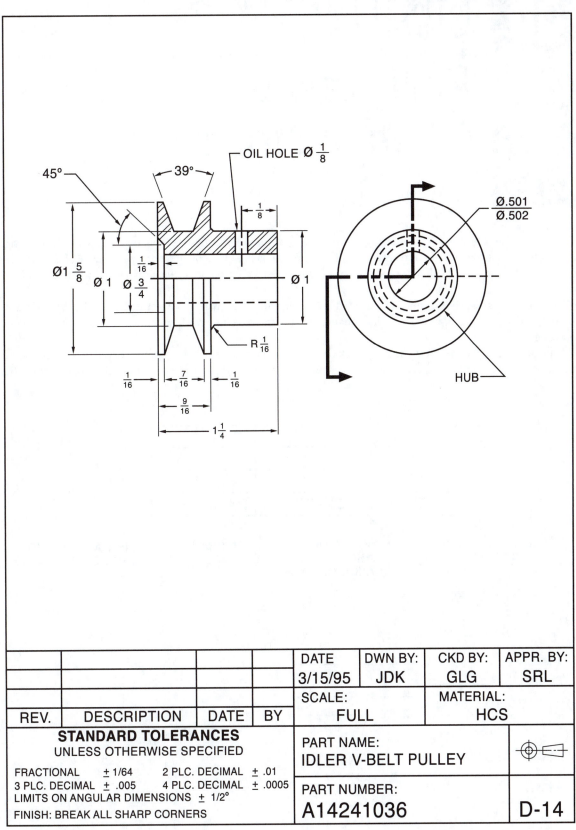

				DATE	DWN BY:	CKD BY:	APPR. BY:
				3/15/95	JDK	GLG	SRL
				SCALE:		MATERIAL:	
REV.	DESCRIPTION	DATE	BY	FULL		HCS	

STANDARD TOLERANCES
UNLESS OTHERWISE SPECIFIED

FRACTIONAL ± 1/64	2 PLC. DECIMAL ± .01
3 PLC. DECIMAL ± .005	4 PLC. DECIMAL ± .0005
LIMITS ON ANGULAR DIMENSIONS ± 1/2°	
FINISH: BREAK ALL SHARP CORNERS	

PART NAME:
IDLER V-BELT PULLEY

PART NUMBER:
A14241036

D-14

16 UNIT

Revisions or Change Notes

Completed drawings often require changes or revisions in the original design. This may be the result of design improvement, customer request, or correction of an original error. Quite often, changes are made after a part has gone into production. Machining processes or efforts to reduce costs often dictate when or where a change is required.

When a change is made, it must be recorded on the drawing. The change, who made it, and the date of change should be documented, Figure 16.1. Minor changes in size are frequently made without altering original lines on a drawing.

The place on a print where a change was made may be indicated by a circled number or letter. The change can be referenced quickly by comparing the numbers or letters on the print and those in the revision box, Figure 16.2.

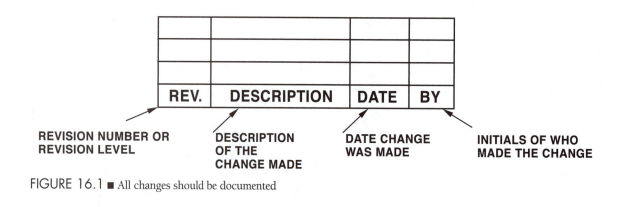

REV.	DESCRIPTION	DATE	BY

REVISION NUMBER OR REVISION LEVEL   DESCRIPTION OF THE CHANGE MADE   DATE CHANGE WAS MADE   INITIALS OF WHO MADE THE CHANGE

FIGURE 16.1 ■ All changes should be documented

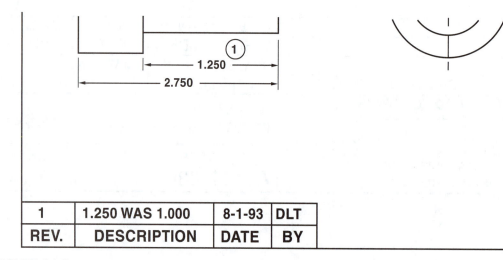

1	1.250 WAS 1.000	8-1-93	DLT
REV.	DESCRIPTION	DATE	BY

FIGURE 16.2 ■ Circled numbers on a print indicate a change

ASSIGNMENT D-15: SLOTTED PLATE

1. How thick is the slotted plate? _____

2. What material is required for the plate? _____

3. What change was made at ①? _____

4. What view of the plate is shown? _____

5. What tolerance is allowed on the 1.00 dimension? _____

6. What tolerance is allowed on the 3.00 dimension? _____

7. On what date was a change made to the drawing? _____

8. Determine the overall length of the slot that is cut in the plate. _____

9. Determine distance Ⓧ. _____

10. Determine distance Ⓨ. _____

11. Determine distance Ⓩ. _____

12. What is the diameter of the hole in the plate? _____

13. What is another name for a change note? _____

14. Why are change notes recorded on a drawing? _____

15. What could have been used in place of the ① to indicate where a change was made? _____

NOTE:
MATERIAL .12 THICK

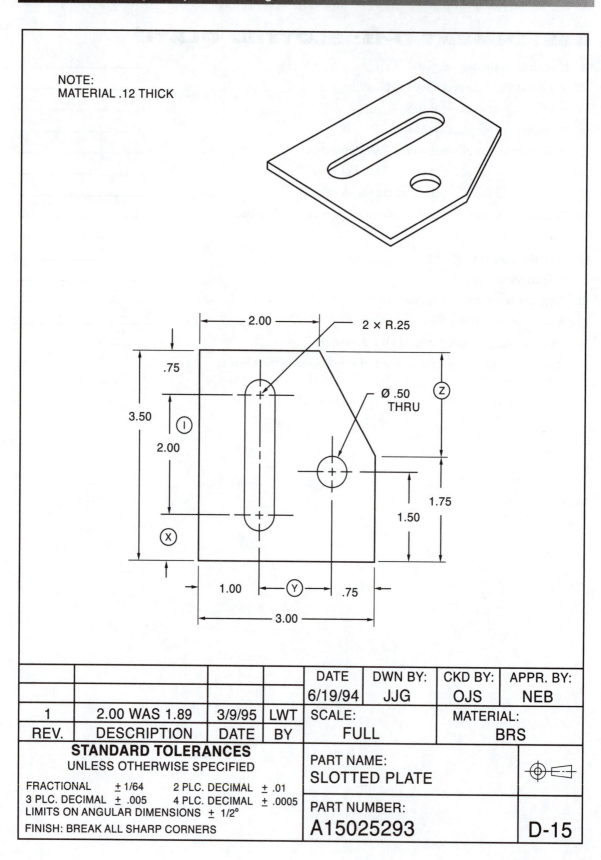

DATE	DWN BY:	CKD BY:	APPR. BY:
6/19/94	JJG	OJS	NEB

1	2.00 WAS 1.89	3/9/95	LWT	SCALE:		MATERIAL:	
REV.	DESCRIPTION	DATE	BY	FULL		BRS	

STANDARD TOLERANCES
UNLESS OTHERWISE SPECIFIED

FRACTIONAL ± 1/64 2 PLC. DECIMAL ± .01
3 PLC. DECIMAL ± .005 4 PLC. DECIMAL ± .0005
LIMITS ON ANGULAR DIMENSIONS ± 1/2°
FINISH: BREAK ALL SHARP CORNERS

PART NAME:
SLOTTED PLATE

PART NUMBER:
A15025293

D-15

Counterbores, Countersinks, and Spotfaces

COUNTERBORES

A counterbored hole is one that has been enlarged at one end to provide clearance for the head of a screw, bolt, or pin. The counterbore is usually machined to a depth that is equal to or slightly more than the thickness of the head on the screw, bolt, or pin. This allows the head to be recessed into the surface of the workpiece, Figure 17.1.

COUNTERSINKS

A countersunk hole has a cone-shaped angle called a countersink cut into one end of the hole. The angle of the countersink is generally 82 degrees to match the angle found on the head of a flathead screw. Countersinks are often slightly larger that the diameter of the screw head. This allows the top of the screw head to be flush or slightly below the surface of the workpiece, Figure 17.2.

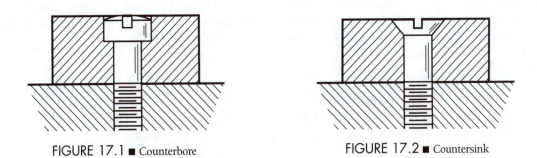

FIGURE 17.1 ■ Counterbore   FIGURE 17.2 ■ Countersink

SPOTFACES

A spotface is similar to a counterbore but usually not as deep, Figure 17.3. The purpose of a spotface is to provide a smooth, flat surface on an irregular surface. A spotface provides a flat bearing surface for a nut, washer, or the head of a screw or bolt.

The diameter and depth of a spotface will vary depending on the size of the nut, washer, or bolt head. Spotfaces are usually machined to a depth of 1/64 to 1/16, depending on the irregularity of the surface being machined.

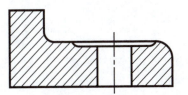

FIGURE 17.3 ■ Spotface

DIMENSIONING COUNTERBORES

Countersunk and counterbored holes are normally dimensioned with a leader line. Dimensions at the end of the leader line specify the minor hole diameter as well as the diameter and depth of the counterbore. The old method of calling out a counterbore is shown in Figure 17.4A. Notes and abbreviations were used to specify the diameter and depth. The new method requires the use of symbols to specify counterbore ⌴ and diameter ∅ and depth ▼ , Figure 17.4B.

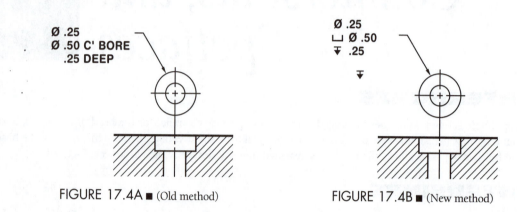

FIGURE 17.4A ■ (Old method)   FIGURE 17.4B ■ (New method)

DIMENSIONING COUNTERSINKS

Dimensions for countersunk holes usually include the minor hole diameter, depth if not a through hole, countersink angle, and the required finished diameter of the countersink. Figures 17.5A and 17.5B show the old method and new method of calling out countersinks. Note that the old method of specifying a countersink was to use the letters CSK. The new method replaces the CSK abbreviation with the symbol ∨ .

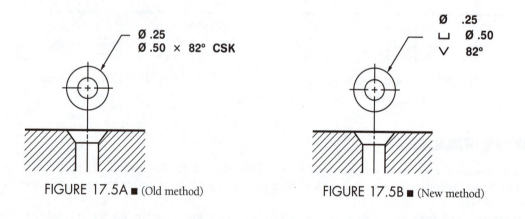

FIGURE 17.5A ■ (Old method)   FIGURE 17.5B ■ (New method)

DIMENSIONING SPOTFACES

The diameter and depth of a spotface is usually specified on the drawing. In some cases, however, the thickness of remaining material is specified. The old method of dimensioning was to use the letters SF to specify a spotface, Figure 17.6A. Here again, symbols are used to replace notes and abbreviations, Figure 17.6B. If dimension is not provided for the depth of the spotface or the thickness of remaining material, the spotface should be cut to the minimum depth necessary to clean up the bearing surface to the specified diameter.

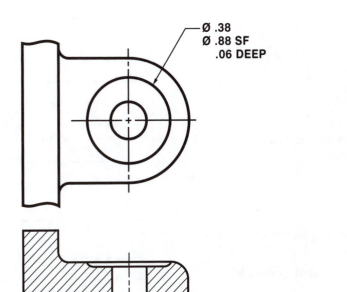

Ø .38
Ø .88 SF
.06 DEEP

FIGURE 17.6A ■ (Old method)

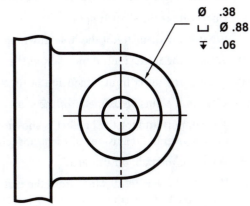

Ø   .38
⊔ Ø .88
⊽ .06

FIGURE 17.6B ■ (New method)

ASSIGNMENT D-16: SHAFT SUPPORT

1. How many counterbored holes are there? _____

2. What is the diameter of the counterbore? _____

3. What is the diameter of the thru hole to be counterbored? _____

4. What is the depth of the counterbore? _____

5. What is the diameter of the oil hole? _____

6. At what angle to the vertical is the oil hole? _____

7. How wide is the shaft support? _____

8. What is the diameter of the hole in which the shaft is to run? _____

9. What is the upper limit dimension of the shaft hole? _____

10. What is the lower limit dimension of the shaft hole? _____

11. What is the outside diameter of the shaft support? _____

12. The support arm for the bracket is shown in section A–A. What are the outside overall dimensions of the arm section? _____

13. What are the radii on this arm? _____

14. What is the vertical distance from the centerline of the slot to the centerline of the shaft? _____

15. What is the horizontal distance from the finished surface on the back of the support to the centerline of the shaft? _____

16. How far from the top of the support is the centerline of the slot? _____

17. What is the width of the slot? _____

18. To what depth is the slot cut into support? _____

19. Determine distance Ⓐ. _____

20. The length of the shaft support has been changed to 2.75. What was the length before the change was made? _____

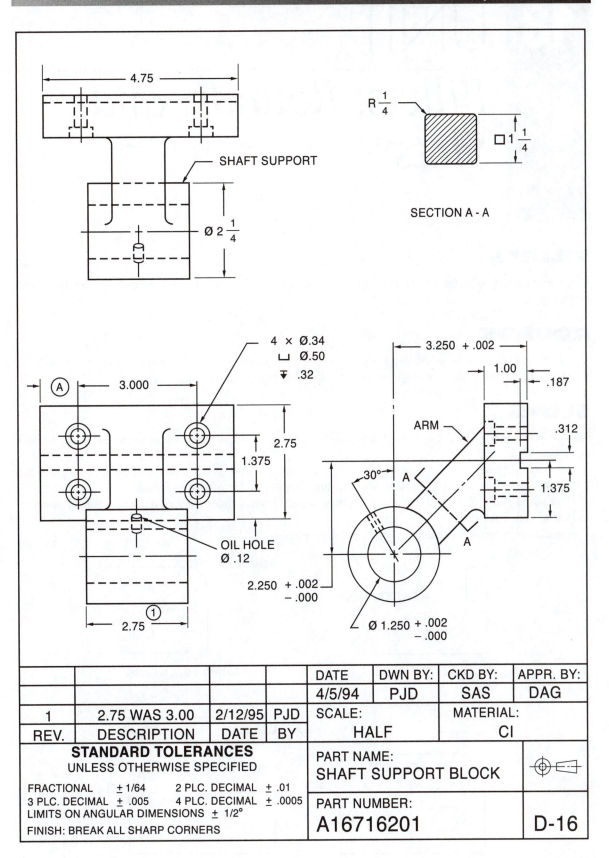

SHAFT SUPPORT

4.75

Ø 2 1/4

R 1/4

□ 1 1/4

SECTION A - A

4 × Ø.34
⊔ Ø.50
▼ .32

A

3.000

2.75

1.375

OIL HOLE
Ø .12

2.75

①

3.250 + .002

1.00

.187

ARM

.312

1.375

30°

A

A

A

2.250 + .002
− .000

Ø 1.250 + .002
− .000

				DATE	DWN BY:	CKD BY:	APPR. BY:
				4/5/94	PJD	SAS	DAG
1	2.75 WAS 3.00	2/12/95	PJD	SCALE:		MATERIAL:	
REV.	DESCRIPTION	DATE	BY	HALF		CI	

STANDARD TOLERANCES
UNLESS OTHERWISE SPECIFIED

FRACTIONAL ± 1/64   2 PLC. DECIMAL ± .01
3 PLC. DECIMAL ± .005   4 PLC. DECIMAL ± .0005
LIMITS ON ANGULAR DIMENSIONS ± 1/2°
FINISH: BREAK ALL SHARP CORNERS

PART NAME:
SHAFT SUPPORT BLOCK

PART NUMBER:
A16716201

D-16

18 UNIT

Fillets, Rounds, and Slots

FILLETS

A *fillet* is additional metal allowed in the inner intersection of two surfaces, Figure 18.1. A fillet increases the strength of the object.

ROUNDS

A *round* is an outside radius added to a piece, Figure 18.2. A round improves the appearance of an object. It also avoids forming a sharp edge that might cause interference or chip off under a sharp blow.

SLOTS

Slots are mainly used on machines to hold parts together. The two principal types of slots are the T-slot and the dovetail, Figure 18.3.

T-slots are frequently used for clamping work securely. Drill press and milling machine tables are examples where T-slots are used. Work to be machined or work-holding devices are clamped using T-slots.

Dovetails are used where two machine parts must move. Dovetail slides are used on lathes, milling machines, and various other machines.

Special cutters are used to machine T-slots and dovetail slots.

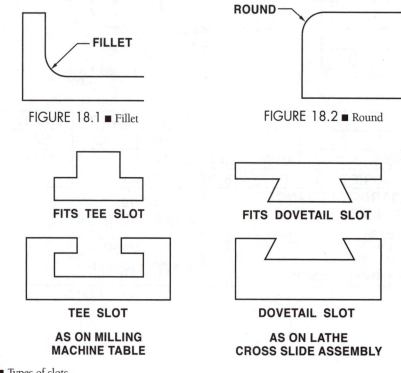

FIGURE 18.1 ■ Fillet

FIGURE 18.2 ■ Round

FITS TEE SLOT

FITS DOVETAIL SLOT

TEE SLOT

DOVETAIL SLOT

AS ON MILLING
MACHINE TABLE

AS ON LATHE
CROSS SLIDE ASSEMBLY

FIGURE 18.3 ■ Types of slots

ASSIGNMENT D-17: COMPOUND REST SLIDE

1. In which view is the shape of the dovetail shown? _____

2. In which view is the shape of the T-slot shown? _____

3. How many rounds are shown in the top view? _____

4. In which view is the fillet shown? _____

5. What line in the top view represents surface Ⓡ of the side view? _____

6. What line in the top view represents surface Ⓛ of the side view? _____

7. What line in the side view represents surface Ⓐ of the top view? _____

8. What is the distance from the base of the slide to line Ⓙ? _____

9. How wide is the opening in the dovetail? _____

10. What two lines in the top view indicate the opening of the dovetail? _____

11. In the side view, how far is the lower left edge of the dovetail from the left side of the piece? _____

12. What is the length of dimension Ⓨ? _____

13. What is the vertical distance from the surface represented by the line Ⓠ to that represented by line Ⓣ? _____

14. What dimension represents the distance between lines Ⓕ and Ⓖ? _____

15. What is the overall depth of the T-slot? _____

16. What is the width of the bottom of the T-slot? _____

17. What is the height of the opening at the bottom of the T-slot? _____

18. What is the length of dimension Ⓥ? _____

19. What is the length of dimension Ⓧ? _____

20. What is the horizontal distance from line Ⓝ to line Ⓢ? _____

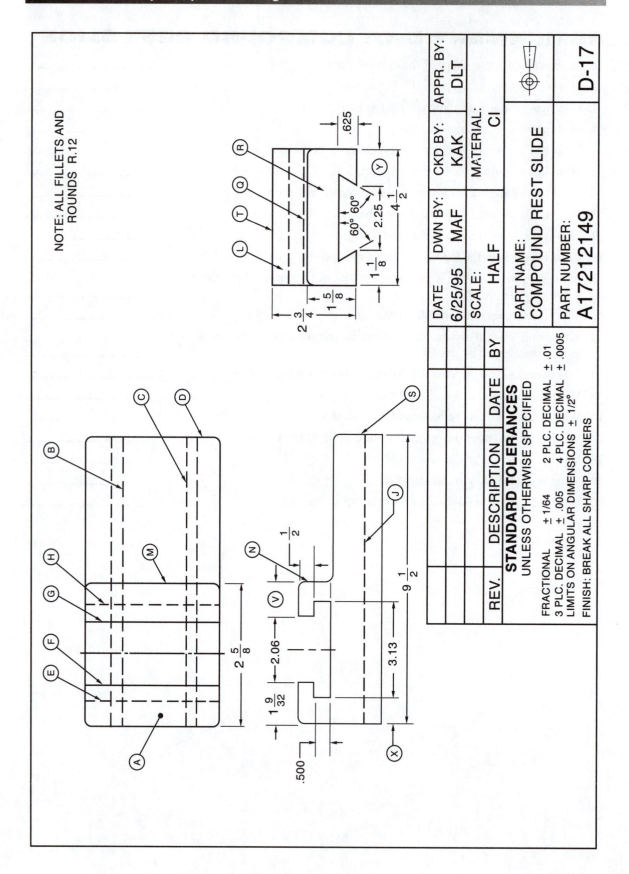

NOTE: ALL FILLETS AND
ROUNDS R.12

	DATE	DWN BY:	CKD BY:	APPR. BY:
	6/25/95	MAF	KAK	DLT
	SCALE:		MATERIAL:	
	HALF		CI	

PART NAME:
COMPOUND REST SLIDE

PART NUMBER:
A1721149

D-17

REV.	DESCRIPTION	DATE	BY

STANDARD TOLERANCES
UNLESS OTHERWISE SPECIFIED

FRACTIONAL ± 1/64   2 PLC. DECIMAL ± .01
3 PLC. DECIMAL ± .005   4 PLC. DECIMAL ± .0005
LIMITS ON ANGULAR DIMENSIONS ± 1/2°
FINISH: BREAK ALL SHARP CORNERS

Machining Symbols, Bosses, and Pads

MACHINING SYMBOLS

Castings or forgings often require specific surfaces to be finished by machining. To illustrate these surfaces, a symbol called a *machining symbol* or finish mark is used.

The American National Standards Institute recommends a standard system of symbols for surface finish. This new system replaces the old V or f symbols formerly used on industrial drawings. Figure 19.1 shows common types of machining symbols.

Machining symbols are placed with the point on the finished surface, Figure 19.2. They are often placed on an extension line or leader. Like dimensions, finish marks should only appear once on a blueprint. They should not be duplicated from one view to another.

Older industrial drawings may specify a desired finish by the use of a series of code numbers or letters to meet their own particular needs. The letter G, for example, is still used either with the $\sqrt{G}$ or alone to denote a surface finished by grinding. In other instances, finished surfaces may be specified by the older symbol (f), Figure 19.3.

BOSSES AND PADS

Bosses and pads serve the same function. They are raised surfaces that are machined to provide a smooth surface for mating parts.

SYMBOL	DEFINITION
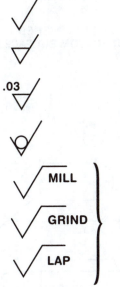	USED TO INDICATE THAT A SURFACE MAY BE PRODUCED BY ANY METHOD.
	SURFACE TO BE FINISHED BY MACHINING. EXTRA MATERIAL MUST BE PROVIDED FOR MACHINING.
.03	INDICATES THE AMOUNT OF MATERIAL TO BE REMOVED BY MACHINING.
	INDICATES THAT MATERIAL REMOVAL IS NOT ALLOWED. USUALLY FOUND ON CAST OR FORGED SURFACES.
MILL / GRIND / LAP	ONLY USED WHEN SPECIFICATION OF THE MACHINING PROCESS IS ESSENTIAL.

FIGURE 19.1 ■ Machining symbols

A *boss* is a round, raised surface of relatively small size above the surface of an object, Figure 19.4A.

A *pad* is a raised surface of any shape except round. This raised surface is above the surface of the object, Figure 19.4B.

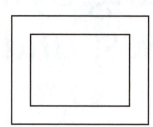
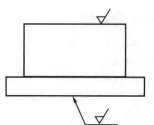

FIGURE 19.2 ■ New style finish marks

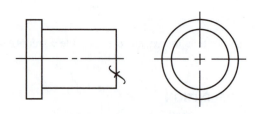

INDICATING A FINISHED FACE

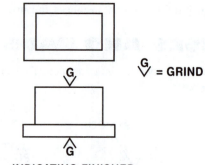

$\overset{G}{\triangledown}$ = GRIND

INDICATING FINISHED TOP AND BOTTOM SURFACES

FIGURE 19.3 ■ Old style finish marks

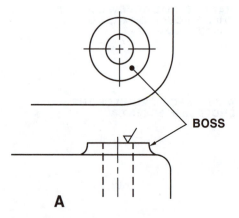

BOSS

A

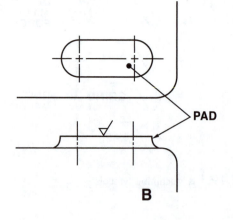

PAD

B

FIGURE 19.4 ■ Bosses and pads

ASSIGNMENT D-18: RACK COLUMN BRACKET

1. How many holes are shown on the drawing? _____

2. What is the height of the base, not including the height of the bosses? _____

3. What is the height of the bosses above the top of the base? _____

4. What is the radius on the corners of the base? _____

5. What is the diameter of the bosses on the base? _____

6. Determine distance Ⓐ. _____

7. Determine distance Ⓑ. _____

8. How far from the horizontal centerline in the top view are the bosses? _____

9. The outside of the pad on the upright support is how far from the centerline of the upright? _____

10. What approximate fractional dimension is the $\varnothing$ .688 $^{+.002}_{-.000}$ hole? _____

11. How far is the centerline of the $\frac{1.002}{1.000}$ diameter hole from the centerline of the $\varnothing$ .688 hole? _____

12. What size is the hole that is parallel to the base? _____

13. What is the distance from the bottom of the base to the center of the $\varnothing$ .688 $^{+.002}_{-.000}$? _____

14. What is the upper limit dimension of the upright hole? _____

15. What is the lower limit dimension of the upright hole? _____

16. What is the outside diameter of the upright? _____

17. How far does the $\varnothing$ .688 $^{+.002}_{-.000}$ hole cut into the $\frac{1.002}{1.000}$ hole? _____

18. How wide is the pad? _____

19. What change was made on the drawing? _____

20. How many $\varnothing$ .41 holes are required? _____

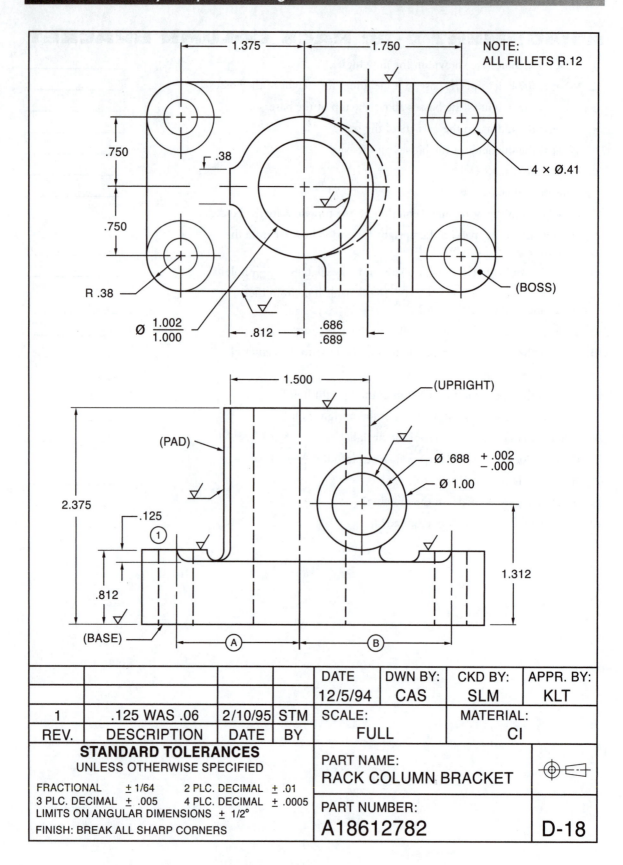

NOTE:
ALL FILLETS R.12

1.375   1.750

.750

.38

.750

4 × Ø.41

R .38

(BOSS)

Ø $\frac{1.002}{1.000}$ .812 $\frac{.686}{.689}$

1.500   (UPRIGHT)

(PAD)

Ø .688 $\begin{array}{c}+.002\\-.000\end{array}$

Ø 1.00

2.375

.125

①

.812

1.312

(BASE)   Ⓐ   Ⓑ

			DATE	DWN BY:	CKD BY:	APPR. BY:
			12/5/94	CAS	SLM	KLT
1	.125 WAS .06	2/10/95	STM	SCALE:		MATERIAL:
REV.	DESCRIPTION	DATE	BY	FULL		CI

STANDARD TOLERANCES
UNLESS OTHERWISE SPECIFIED

FRACTIONAL ± 1/64   2 PLC. DECIMAL ± .01
3 PLC. DECIMAL ± .005   4 PLC. DECIMAL ± .0005
LIMITS ON ANGULAR DIMENSIONS ± 1/2°
FINISH: BREAK ALL SHARP CORNERS

PART NAME:
RACK COLUMN BRACKET

PART NUMBER:
A18612782

D-18

Tapers, Chamfers, and Bevels

TAPERS

A taper is defined as a gradual and uniform increase or decrease in size along a given length of a part. Tapers may be conical or flat and are specified on a drawing in degrees, taper per foot, taper per inch, as a standard taper, or as a ratio. Taper per foot and taper per inch refers to the required variation in size along one inch or one foot of taper length.

Conical Tapers

A tapered surface on a round part is called a conical taper. Internal and external conical tapers are used extensively for alignment and holding purposes between mating parts. Machine spindles and tapered shank tools such as drills, reamers, mill cutters, and lathe centers, for example, have standard tapers. Standard tapers may be specified on a drawing by taper name and number, Figure 20.1.

Tapers that are nonstandard are usually specified in degrees or dimensioned by giving a diameter at one end of the taper, the length of taper, and the taper per inch or taper per foot, Figure 20.2.

A taper may also be shown as a ratio of the difference in diameters. When conical tapers are specified as a ratio, a conical taper symbol ▷── is used, Figure 20.3.

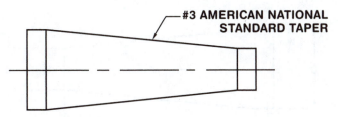

FIGURE 20.1 ■ Standard taper

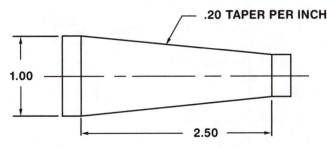

FIGURE 20.2 ■ Taper per inch

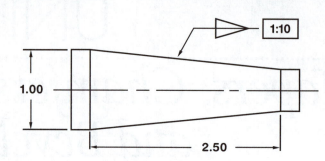

FIGURE 20.3 ■ Conical taper as a ratio

Determining Taper Per Inch and Taper Per Foot

If the taper per inch, taper per foot, or slope is not specified on the drawing, it can be determined if the large diameter, small diameter, and length of taper are known. The taper per inch or TPI can be determined by subtracting the diameter at the small end of the taper (d) from the diameter at the large end of the taper (D) and dividing the result by the length of the taper (L), Figure 20.4. The resulting formula is as follows: $\text{TPI} = (D - d)/L$. To determine the taper per foot, multiply the taper per inch by twelve.

Flat Tapers

Flat tapers are defined as a slope or inclined surface on a flat object. A flat taper may be specified in degrees or as a ratio of the difference in heights at each end of the taper. When specified as a ratio, a slope symbol ▷ is used, Figure 20.5.

Determining Slope

The slope of a flat taper can be determined by subtracting the height at the small end of the taper (h) from the height at the large end of the taper (H) and dividing the result by the length of the taper (L), Figure 20.6. The resulting formula is as follows: $\text{Slope} = (H - h)/L$.

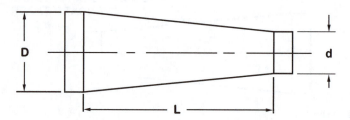

FIGURE 20.4 ■ Taper determined by dimensions

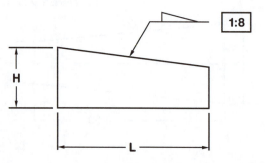

FIGURE 20.5 ■ Flat taper as a ratio

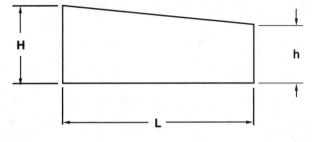

FIGURE 20.6 ■ Slope = (H–h)/L

CHAMFERS

A *chamfer* is an angle cut on the end of a shaft or on the edge of a hole. Chamfers remove sharp edges, add to the appearance of the job, and provide handling safety. Chamfers also enable parts to be assembled more easily. The ends of screws and bolts are chamfered for this reason.

Chamfered edges are usually cut at included angles from 30 degrees to 90 degrees. The dimension for a chamfer may be shown as an angle cut to the axis, or centerline, of the part as at Figure 20.7A or as a total included angle as shown at 20.7C. The length of the chamfer may be specified as a linear length as at Figure 20.7A or as a diameter as shown in Figures 20.7B and 20.7C.

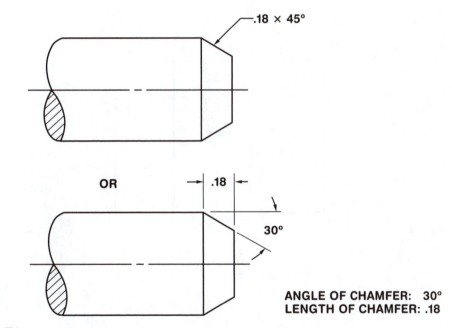

FIGURE 20.7A ■ External chamfer

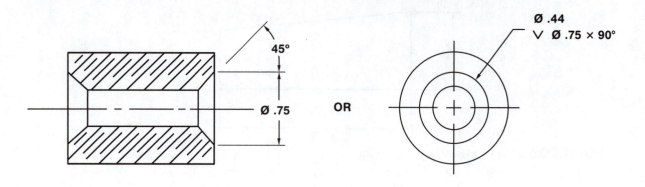

FIGURE 20.7B AND FIGURE 20.7C ■ Internal chamfer

BEVELS

A *bevel,* or a beveled surface, is a cut at an angle to some horizontal or vertical surface or to the axis of the piece. The bevel runs the entire length or width of the piece, Figure 20.8.

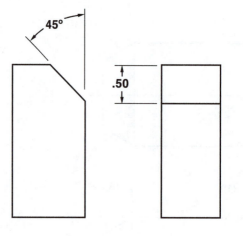

FIGURE 20.8 ■ Bevel

ASSIGNMENT D-19: OFFSET BRACKET

1. What is the height of the shaft carrier? _____

2. At what angle is the offset arm to the body of the piece? _____

3. What is the center-to-center measurement of the length of the offset arm? _____

4. What radius forms the upper end of the offset arm? _____

5. What is the width of the bolt slot in the body of the bracket? _____

6. What is the length, center-to-center, of this slot? _____

7. What is the overall width of the pad? _____

8. What is the radius of the fillet between this pad and the edge of the piece? _____

9. What size chamfer is required on the ∅ .375 hole? _____

10. How thick are the body and the pad together? _____

11. What size oil hole is required in the shaft carrier? _____

12. How far is the oil hole from the finished face on the shaft carrier? _____

13. What tolerance is applied to the ∅ .375 hole? _____

14. What is the diameter of the shaft carrier body? _____

15. What is the distance from the finished face on the shaft carrier to the finished face on the pad? _____

16. How much is the distance Ⓑ? _____

17. How much is the distance Ⓐ? _____

18. What term is used to indicate the uniform change in size of the offset arm? _____

19. If 1/2-inch bolts are used in holding the bracket to the machine base, what is the total clearance on the sides of the slot? _____

20. If the center-to-center distance of two 1/2-inch bolts that fit in the slot is 1.50 inches, what is the total clearance on the ends of the slot? _____

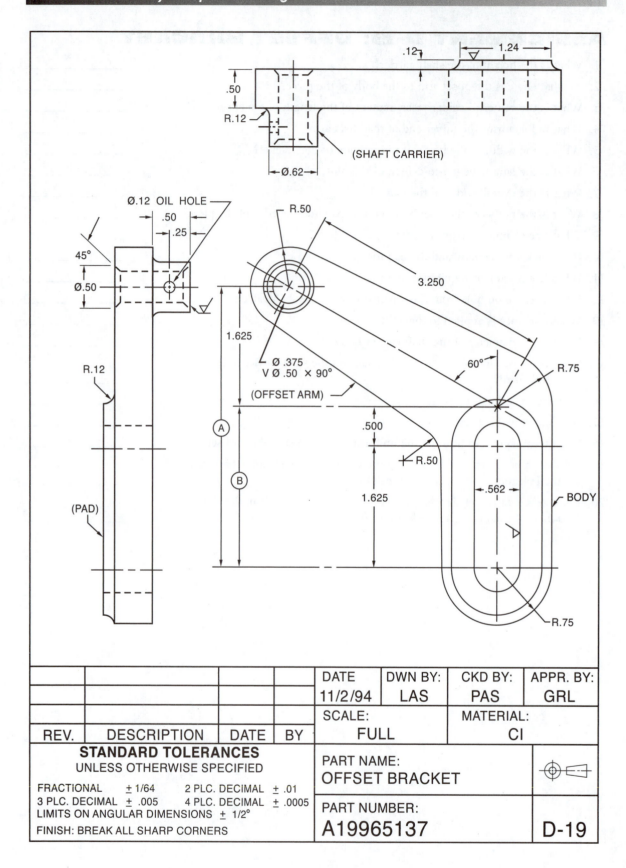

.12

1.24

.50

R.12

Ø.62

(SHAFT CARRIER)

Ø.12 OIL HOLE

.50

.25

45°

Ø.50

R.50

3.250

1.625

R.12

Ø .375
V Ø .50 × 90°

(OFFSET ARM)

60°

R.75

Ⓐ

.500

R.50

Ⓑ

.562

BODY

(PAD)

1.625

R.75

				DATE	DWN BY:	CKD BY:	APPR. BY:
				11/2/94	LAS	PAS	GRL

				SCALE:		MATERIAL:	
REV.	DESCRIPTION	DATE	BY	FULL		CI	

STANDARD TOLERANCES
UNLESS OTHERWISE SPECIFIED

FRACTIONAL ± 1/64   2 PLC. DECIMAL ± .01
3 PLC. DECIMAL ± .005   4 PLC. DECIMAL ± .0005
LIMITS ON ANGULAR DIMENSIONS ± 1/2°
FINISH: BREAK ALL SHARP CORNERS

PART NAME:
OFFSET BRACKET

PART NUMBER:
A19965137

D-19

Necks and Knurling

NECKS AND UNDERCUTS

Necking, or undercutting or grooving as it is sometimes called, is the process of cutting a recess in a diameter. Necks are often cut at the ends of threads or where a shaft changes diameter. The neck undercuts the shaft so that a mating part will seat against the shoulder.

Necks are usually dimensioned with a leader and a note. The width and depth of the recess is provided, Figure 21.1. Another common practice is to call out the neck width and the diameter of the shaft at the bottom of the neck, Figure 21.2.

KNURLING

Knurling is the process of impressing a straight or diamond-shaped pattern into a cylindrical piece using special knurling tools. The knurl is formed by forcing the hardened knurling rollers on the knurling tool into the surface of a revolving cylindrical part. The pressure of the knurling tool creates a pattern of straight or diamond grooves as material is forced outward against the knurling rollers, Figure 21.3.

In addition to creating a pattern on the surface of the cylindrical workpiece, the displacement of metal during the knurling process tends to increase the diameter of the knurled part.

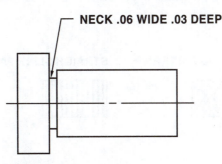

NECK .06 WIDE .03 DEEP

FIGURE 21.1 ■ Dimensioning a neck by width and depth

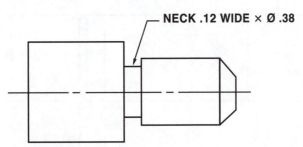

NECK .12 WIDE × Ø .38

FIGURE 21.2 ■ Dimensioning a neck by width and diameter

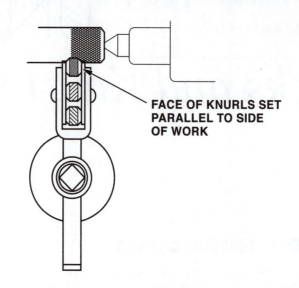

FACE OF KNURLS SET PARALLEL TO SIDE OF WORK

FIGURE 21.3 ■ Knurling

Knurls are made in diamond or straight patterns, Figure 21.4. The diametral pitch of the knurling tool determines the size of the finished knurl. The distance between the knurling grooves decreases as the diametral pitch increases. A 64DP knurl would be coarser than a 128DP knurl, for example.

The most commonly used diametral pitches for knurling are 64DP, 96DP, 128DP, and 160DP. Knurling operations are often performed to improve the appearance of a part, provide a gripping surface, or to increase the diameter of a part when a press fit is required between mating parts. Straight knurls, for example, are often specified to create a tight fit between a shaft and a hole of equal diameter.

Dimensioning Knurls

Knurling is generally dimensioned by specifying type, pitch, length of knurl, and diameter of the part before knurling, Figure 21.5. The finished diameter of the part after knurling may be specified if it is important to do so. If a straight knurling operation is being performed to provide a press fit, for example, the minimum diameter after knurling should be specified.

DIAMOND PATTERN   **STRAIGHT-LINE PATTERN**

FIGURE 21.4 ■ Patterns of knurls

**96 DP
RAISED DIAMOND KNURL**

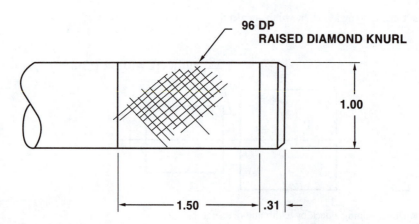

1.00

1.50   .31

FIGURE 21.5 ■ Dimensioning a knurl

ASSIGNMENT D-20: CAM CARRIER SUPPORT

1. What is the largest diameter of the cam carrier support? _____

2. What is the overall length? _____

3. What is the outside diameter of the hub? _____

4. How long is the hub? _____

5. How thick is the flange? _____

6. How many countersunk holes are in the cam carrier support? _____

7. What is the diameter of the circle on which the countersunk holes are located? _____

8. How many degrees apart are the countersunk holes spaced? _____

9. What size screw must fit the countersunk holes? _____

10. What specifications are called out on the neck? _____

11. What size knurl is required? _____

12. What pitch would the knurl be? _____

13. What type of section view is shown? _____

14. What material is specified in the title block? _____

15. What tolerance is permitted on three-place decimal dimensions? _____

16. What is the largest size to which the hole through the center can be machined? _____

17. What is the smallest size to which the hole through the center can be machined? _____

18. What is the diameter of the recess into the ⌀ 2.750? _____

19. What is the depth of the recess? _____

20. What is the amount and degree of chamfer on the ⌀ 2.750? _____

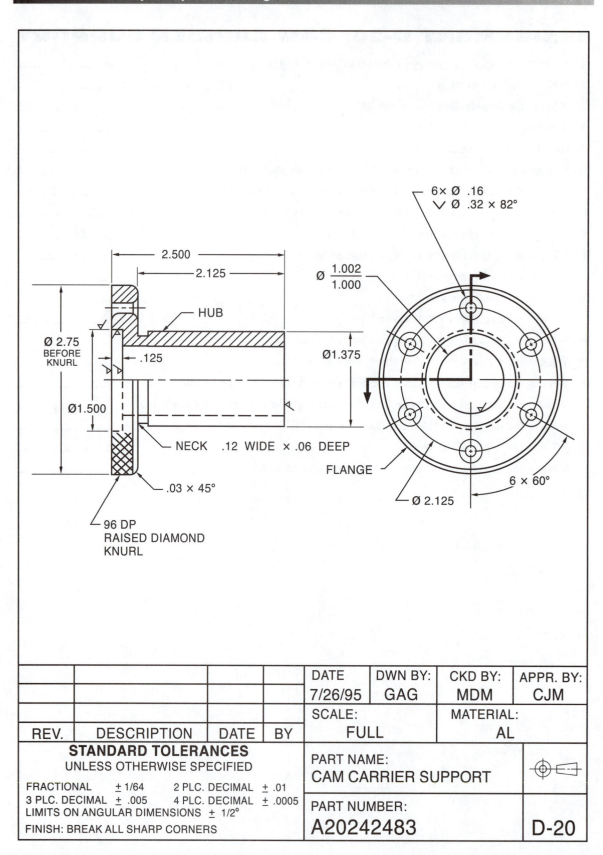

				DATE 7/26/95	DWN BY: GAG	CKD BY: MDM	APPR. BY: CJM
				SCALE: FULL		MATERIAL: AL	
REV.	DESCRIPTION	DATE	BY				

STANDARD TOLERANCES
UNLESS OTHERWISE SPECIFIED

| FRACTIONAL ± 1/64 | 2 PLC. DECIMAL ± .01 |
| 3 PLC. DECIMAL ± .005 | 4 PLC. DECIMAL ± .0005 |
LIMITS ON ANGULAR DIMENSIONS ± 1/2°
FINISH: BREAK ALL SHARP CORNERS

PART NAME:
CAM CARRIER SUPPORT

PART NUMBER:
A20242483

D-20

UNIT **22**

Keyseats and Flats

KEYS AND KEYSEATS

A *key* is a specially shaped piece of metal used to align mating parts or keep parts from rotating on a shaft. Keys are usually standard items available in various sizes.

A *keyseat* or *keyway* is a slot that is cut so the key fits into it. The slot is cut into both mating parts. Figure 22.1 shows various shaped keys and keyseats.

The dimension for a keyseat specifies the width, location, and sometimes the length. Woodruff key sizes are specified by a number.

Dimensioning Keyseats

Keyseats are dimensioned by width, depth, location, and, if necessary, length. Common practice is to dimension the depth of the keyseat from the opposite side of the shaft or hole, Figure 22.2.

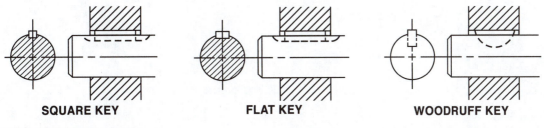

SQUARE KEY   **FLAT KEY**   **WOODRUFF KEY**

FIGURE 22.1 ■ Keyseats and keyways

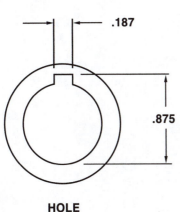

HOLE

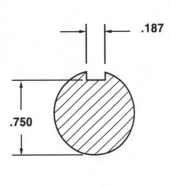

SHAFT

FIGURE 22.2 ■ Dimensioning keyseats

FLATS

Flats are usually cut on rounded or rough surfaces such as shafts or castings, Figure 22.3. Flats provide a surface on which the end of a setscrew can rest when holding an object in place. They also may be provided to fit the jaws of a wrench so a shaft may be turned, Figure 22.4.

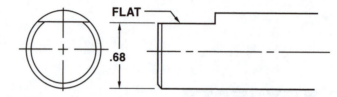

FIGURE 22.3 ■ Specifying a flat in a note

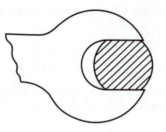

FIGURE 22.4 ■ Flats are provided here to fit the jaws of a wrench

ASSIGNMENT D-21: SPINDLE SHAFT

1. Starting at the bottom end of the shaft and including the neck, what are the successive diameters up to the ⌀ 2.12? _____

2. Starting at the top end of the shaft, what are the successive diameters down to the ⌀ 2.12? _____

3. What is the length of the ⌀ 1.125? _____

4. What is the upper tolerance of size of the ⌀ 1.125? _____

5. What is the lower tolerance of size of the ⌀ 1.125? _____

6. What is the largest size to which the ⌀ 1.250 portion of the shaft can be turned? _____

7. What is the smallest size to which the ⌀ 1.250 portion can be turned? _____

8. How far is it from the bottom end of the shaft to the shoulder of the ⌀ 2.12? _____

9. The cut across the shaft at the point the flat is milled is shown in section A–A. How deep is the flat cut? _____

10. What width and depth is specified for the neck? _____

11. What is the thickness of the ⌀ 2.12 collar? _____

12. How far from the top end of the shaft is the ⌀ 2.12 shoulder? _____

13. How long is the ⌀ 1.500? _____

14. How long is the ⌀ 1.000? _____

15. What is the width and depth of the keyseat? _____

16. What is the length of the keyseat? _____

17. How far is the keyseat from the shoulder of the ⌀ 1.000 $^{+.000}_{-.002}$? _____

18. What is the largest diameter to which the ⌀ 1.500 can be turned? _____

19. What is the amount of chamfer on each end of the piece? _____

20. What revision was made at ②? _____

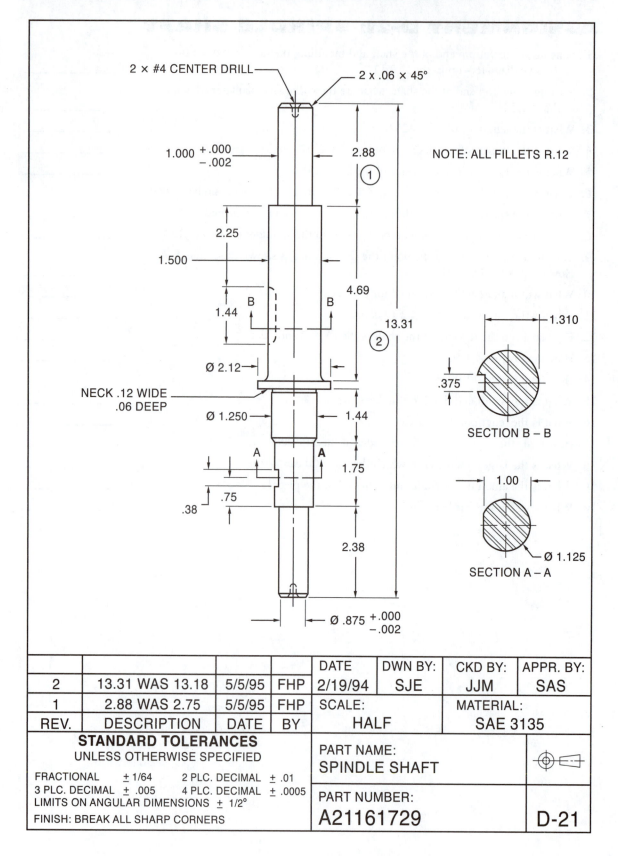

2 × #4 CENTER DRILL

2 x .06 × 45°

1.000 +.000 / −.002

2.88

NOTE: ALL FILLETS R.12

①

2.25

1.500

4.69

B   B

1.44

13.31

②

Ø 2.12

NECK .12 WIDE
.06 DEEP

Ø 1.250

1.44

A   **A**

1.75

.38   .75

2.38

Ø .875 +.000 / −.002

1.310

.375

SECTION B – B

1.00

Ø 1.125

SECTION A – A

2	13.31 WAS 13.18	5/5/95	FHP	DATE 2/19/94	DWN BY: SJE	CKD BY: JJM	APPR. BY: SAS
1	2.88 WAS 2.75	5/5/95	FHP	SCALE: HALF		MATERIAL: SAE 3135	
REV.	DESCRIPTION	DATE	BY				

STANDARD TOLERANCES UNLESS OTHERWISE SPECIFIED	PART NAME: SPINDLE SHAFT	
FRACTIONAL ± 1/64 2 PLC. DECIMAL ± .01 3 PLC. DECIMAL ± .005 4 PLC. DECIMAL ± .0005 LIMITS ON ANGULAR DIMENSIONS ± 1/2° FINISH: BREAK ALL SHARP CORNERS	PART NUMBER: A21161729	D-21

Screw Thread Specification

Screw threads are used in a variety of mechanical applications. They may be used to transfer motion, transmit power, or to fasten parts together. Threaded fasteners such as nuts, bolts, and screws are perhaps the most common application of screw threads.

Screw threads may be internal or external and are available in a variety of diameters, shapes, fits, etc. Therefore, all information needed to fully understand the thread requirements must be specified on the drawing.

SYMBOLS FOR IDENTIFYING THREAD SPECIFICATION

A system of letters and numbers is used to identify thread specification on a print. The sequence used states

■ the major or outside diameter of the thread.

■ the number of threads per inch.

■ the thread form.

■ the thread series.

■ the class of fit.

■ the internal thread or external thread (A=external; B=internal).

Thread identification symbols are shown in Figure 23.1.

Threads may be cut as either a right-hand thread or a left-hand thread. When a right-hand thread is specified, no special symbol is put on the drawing. When a left-hand thread is specified, the symbol LH is placed after the designation of the thread.

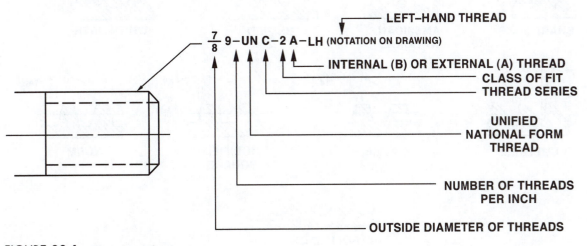

FIGURE 23.1 ■ Thread specification

THREAD FORM

Thread form refers to the shape of the screw thread required. Different thread forms may be specified depending on the application of the thread. Figure 23.2 shows some of the most common thread forms. Note that most thread forms are some variation of a v-shape.

The Unified form is specified for most standard internal and external thread applications where fastening parts together is required.

THREAD SERIES

The *thread series* is the coarseness or fineness of the thread. Unified National threads are commonly called out as coarse (UNC), or fine (UNF), and extra-fine (UNEF) series.

Unified National fine and extra-fine series have more threads per inch for the same diameter than the national coarse series.

UNIFIED NATIONAL THREAD SERIES

The Unified Screw Thread Series was adopted by the United Kingdom, Canada, and the United States. This series establishes screw thread interchangeability among these nations.

The Unified Screw Thread Series is marked as follows:

■ UNC for coarse thread series

■ UNF for fine thread series

■ UNEF for extra-fine thread series

The 8-thread series, marked 8UN, is a uniform-pitch series for large diameters. Uniform pitch means that the number of threads per inch remains constant. For example, an 8-thread series or 8UN thread will have eight threads per inch regardless of the diameter of the part. The same is true of the 12 and 16 uniform pitch series of threads.

The 12-thread series, marked 12UN, is a uniform-pitch series for large diameters requiring threads of a medium-fine pitch.

The 16-thread series, marked 16UN, is a uniform-pitch series for large diameters requiring fine-pitch threads.

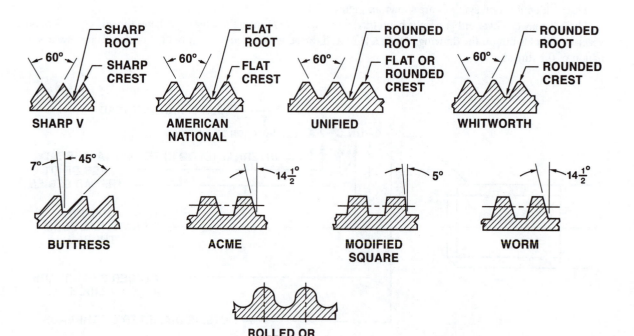

FIGURE 23.2 ■ Most common standard thread forms

CLASSIFICATION OF FIT

Five distinct classes of screw thread fit have been established. This is to ensure interchangeable manufacture of screw thread parts throughout the country.

Class 1: Recommended only for screw thread work where clearance between mating parts is essential for rapid assembly and where shake or play is allowed.

Class 2: Represents a high quality of commercial screw thread product. It is recommended for the great bulk of interchangeable screw thread work.

Class 3: Represents an exceptionally high grade of commercially threaded product. It is recommended only in cases where the high cost of precision tools and continual checking of tools and product is warranted.

Class 4: Intended to meet very unusual requirements more exacting than those for which class 3 is intended. It is a selective fit if initial assembly by hand is required. It is not, as yet, adaptable to quantity production.

Class 5: Includes interchangeable screw thread work consisting of steel studs set in hard materials (cast iron, steel, bronze, etc.) where a wrench-tight fit is required.

CHAMFERING AND NECKING A THREAD DIAMETER

The amount of chamfer or neck on a thread diameter is not always supplied on a working drawing. In this case, the size of chamfer and neck is computed from the thread diameter and number of threads per inch, Figure 23.3.

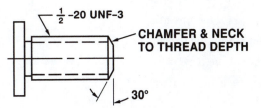

FIGURE 23.3 ■ Indicating chamfer or neck on a thread diameter

ASSIGNMENT D-22: GEAR CARRIER SCREW

1. What is the diameter of the thread? _____

2. What is the number of threads per inch? _____

3. What series of thread is being used? _____

4. How long is that portion on which the thread is cut? _____

5. What thread form is specified? _____

6. How deep is the neck cut? _____

7. What is the length of the ∅ 1.250? _____

8. What is the diameter of the head of the screw? _____

9. How thick is the head of the screw? _____

10. What size chamfer is required on the head of the gear carrier screw? _____

11. What is the upper limit allowed on the width of the neck? _____

12. To what angle is the threaded end of the screw chamfered? _____

13. How wide is the screw slot? _____

14. How deep is the screw slot? _____

15. What is the tolerance on the ∅ 1.250? _____

16. What does the "A" specify on the thread callout? _____

17. What class of fit is required on the thread? _____

18. How deep is the chamfer cut on the thread end? _____

19. What is the overall length of the part? _____

20. What changes have been made on the part? _____

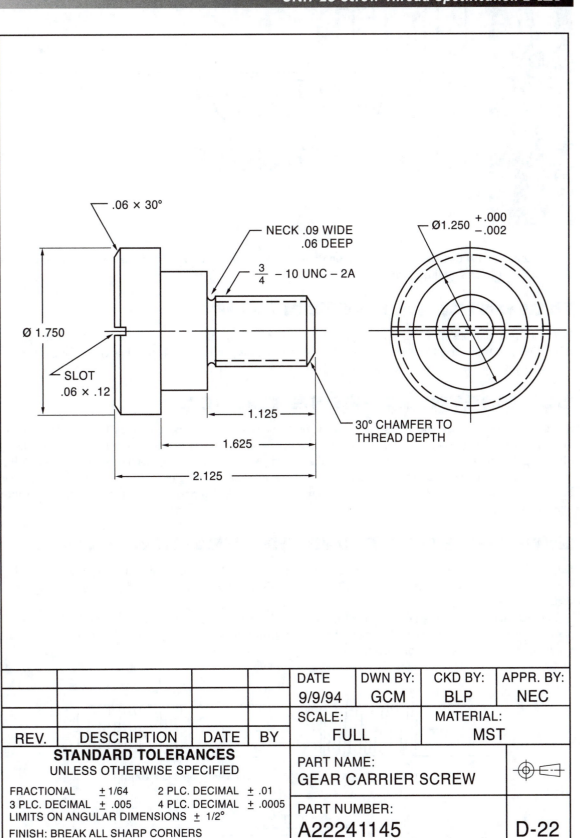

.06 × 30°

NECK .09 WIDE
.06 DEEP

$\frac{3}{4}$ – 10 UNC – 2A

Ø1.250 +.000 −.002

Ø 1.750

SLOT
.06 × .12

1.125

1.625

2.125

30° CHAMFER TO
THREAD DEPTH

DATE	DWN BY:	CKD BY:	APPR. BY:
9/9/94	GCM	BLP	NEC

	SCALE:		MATERIAL:
	FULL		MST

REV.	DESCRIPTION	DATE	BY

STANDARD TOLERANCES
UNLESS OTHERWISE SPECIFIED

FRACTIONAL ± 1/64 2 PLC. DECIMAL ± .01
3 PLC. DECIMAL ± .005 4 PLC. DECIMAL ± .0005
LIMITS ON ANGULAR DIMENSIONS ± 1/2°
FINISH: BREAK ALL SHARP CORNERS

PART NAME:
GEAR CARRIER SCREW

PART NUMBER:
A22241145

D-22

24 UNIT

Screw Thread Representation

Internal and external threads are represented on drawings in one of three ways. They may be drawn pictorially in schematic or using a simplified form.

PICTORIAL REPRESENTATION

Pictorial representations show the thread form very close to how it actually appears, Figure 24.1. The detailed shape is very pleasing to the eye. However, the task of drawing pictorials is very difficult and time consuming. Therefore, pictorial representations are rarely used on threads of less than one inch in diameter.

SCHEMATIC REPRESENTATION

Schematic thread representation does not show the outline of the thread shape, Figure 24.1. Instead, two parallel lines are drawn at the major diameter. The crest and root lines are drawn at right angles to the thread axis instead of sloping. The root lines are drawn thicker than the crest lines. In actual drafting practice, the crest and root lines are spaced by eye to the approximate pitch, and they may be of equal width if preferred. The schematic thread symbol for threads in section does not have the 60-degree thread outline. On one edge the thread outline is advanced one-half of the pitch.

SIMPLIFIED THREAD REPRESENTATION

Simplified thread symbols are used to further reduce drafting time. The thread outline and the crest and root lines are not drawn. Two dotted lines are drawn parallel to the axis, to indicate the depth and the length of the threads, Figure 24.1.

Internal simplified threads are represented the same as the schematic representations, Figure 24.2. In a section view, however, a hidden line is used to show the crest, or major diameter, of the thread. An object line is used to show the root, minor diameter, of the thread.

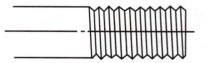

PICTORIAL REPRESENTATION

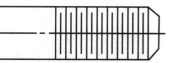

SCHEMATIC REPRESENTATION

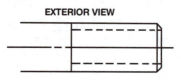

EXTERIOR VIEW

SIMPLIFIED REPRESENTATION

FIGURE 24.1 ■ Thread representation

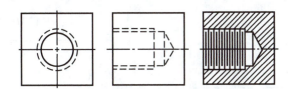

SCHEMATIC REPRESENTATION

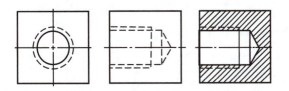

SIMPLIFIED REPRESENTATION

FIGURE 24.2 ■ Representation of internal threads

REPRESENTING TAPPED HOLES

Tapped or *threaded holes* may go all the way through a piece or only part way. Holes that are tapped through are represented as shown in Figure 24.3.

Threads that are not tapped through are shown in combination with the tap drill hole. Tap drill holes are represented with hidden lines showing the diameter of the drill. This same surface is used to represent the minor diameter of the screw thread. The bottom of the tap drill hole is pointed to represent the drill point, Figure 24.4.

Tapped holes may be shown threaded to the bottom of the drill hole or to a specified depth. The specified depth is called out on the print, Figure 24.5.

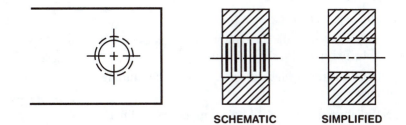

SCHEMATIC   **SIMPLIFIED**

FIGURE 24.3 ■ Hole that is tapped through

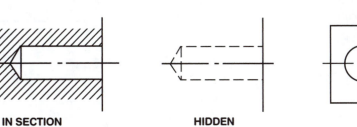

IN SECTION   **HIDDEN**

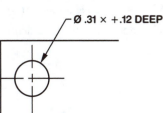

Ø .31 × +.12 DEEP

FIGURE 24.4 ■ Representing tap drill holes

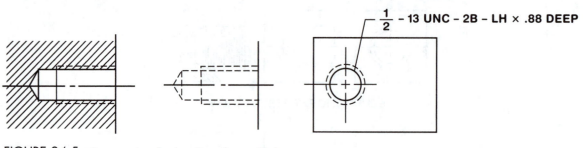

FIGURE 24.5 ■ Representing the threading of tapped holes

ASSIGNMENT D-23: STUFFING BOX

1. What type of thread representation is used to show the 1 3/4–16 thread? _____

2. What is the thread diameter of the tapped holes? _____

3. What is the number of threads per inch for the tapped holes? _____

4. What thread series is used? _____

5. What size thread is cut on the outside of the piece? _____

6. What type of thread repesentation is used on the tapped holes? _____

7. What is the length of the outside thread including the chamfer? _____

8. How much clearance is between the last thread and the flange? _____

9. What is the fillet size between the thread diameter and the flange? _____

10. How thick is the flange? _____

11. What is the overall length of the flange? _____

12. What is the overall width of the flange? _____

13. How far apart are the tapped holes? _____

14. What is the tolerance allowed on the dimension that specifies how far apart the tapped holes are spaced? _____

15. What is the smallest diameter to which the hole through the stuffing box can be machined? _____

16. What is the largest diameter to which the hole through the stuffing box can be machined? _____

17. What height is the ∅ 2.000? _____

18. What tolerance is allowed on the height of the ∅ 2.000? _____

19. What is the angle and depth of chamfer on the ∅ 1.125 hole? _____

20. What does the $\sqrt{\text{FILE}}$ indicate? _____

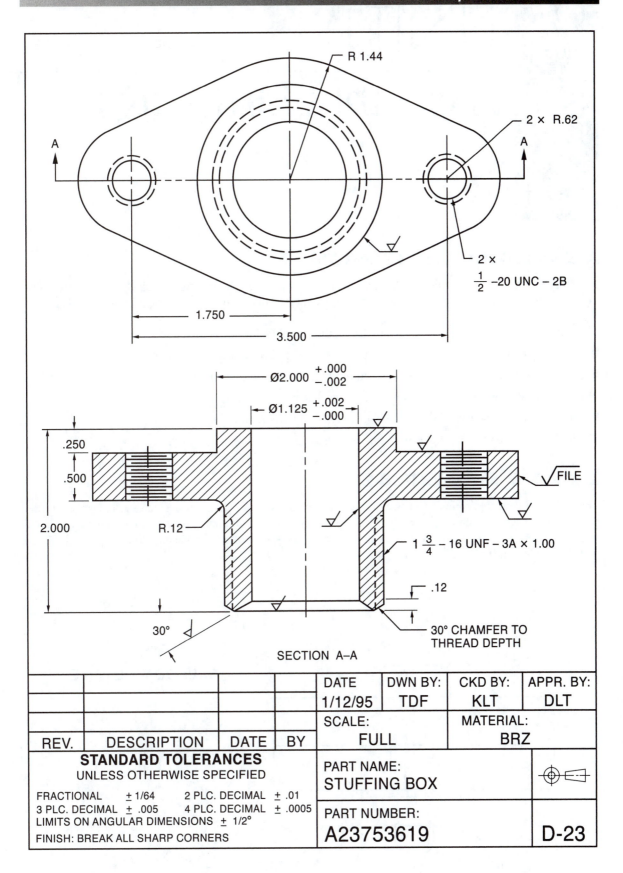

R 1.44

2 × R.62

A — A

2 ×
$\frac{1}{2}$ –20 UNC – 2B

1.750

3.500

Ø2.000 $^{+.000}_{-.002}$

Ø1.125 $^{+.002}_{-.000}$

.250

.500

2.000

R.12

FILE

2.000

$1\frac{3}{4}$ – 16 UNF – 3A × 1.00

.12

30°

30° CHAMFER TO
THREAD DEPTH

SECTION A–A

				DATE	DWN BY:	CKD BY:	APPR. BY:
				1/12/95	TDF	KLT	DLT
				SCALE:		MATERIAL:	
REV.	DESCRIPTION	DATE	BY	FULL		BRZ	

STANDARD TOLERANCES
UNLESS OTHERWISE SPECIFIED

FRACTIONAL ± 1/64	2 PLC. DECIMAL ± .01	
3 PLC. DECIMAL ± .005	4 PLC. DECIMAL ± .0005	
LIMITS ON ANGULAR DIMENSIONS ± 1/2°		
FINISH: BREAK ALL SHARP CORNERS		

PART NAME:
STUFFING BOX

PART NUMBER:
A23753619

D-23

Assembly Drawings

ASSEMBLY DRAWINGS

Industrial drawings often show two or more parts that must be put together to form an *assembly*. An assembly drawing shows the parts or details of a machine or structure in their relative positions as they appear in a completed unit, Figure 25.1.

In addition to showing how the parts fit together, the assembly drawing is used to

■ represent the proper working relationship of the mating parts and the function of each.

■ provide a visual image of how the finished product should look when assembled.

■ provide overall assembly dimensions and center distances.

■ provide a bill of materials for machined or purchased parts required in the assembly.

■ supply illustrations that may be used for catalogs.

DETAIL DRAWINGS

The individual parts that comprise an assembly are referred to as *details*. These details may be standard purchased parts such as machine screws, bolts, washers, springs, and so on, or nonstandard parts that must be manufactured. Unaltered purchased parts do not require a detail drawing. The specifications for standard units are provided in the parts list or bill of materials.

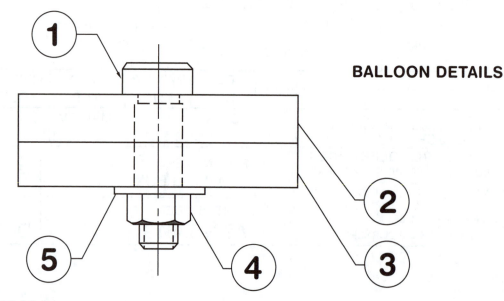

BALLOON DETAILS

FIGURE 25.1 ■ Assembly drawing

Nonstandard parts require drawings that may appear on one sheet or on separate sheets. The detail drawings supply more specific information than is provided on the assembly drawing. All views, dimensions, and notes required to describe the part completely appear on the detail drawing, Figure 25.2.

IDENTIFICATION SYMBOLS

The details of a mechanism are identified on an assembly with reference letters or numbers. These letters or numbers are contained in circles, or balloons, with leaders running to the part to which each refers, Figure 25.3. These symbols are also included in a parts list that gives a descriptive title for each part.

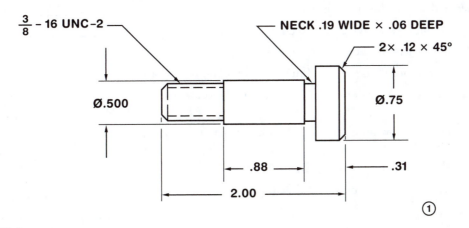

FIGURE 25.2 ■ Detail drawing

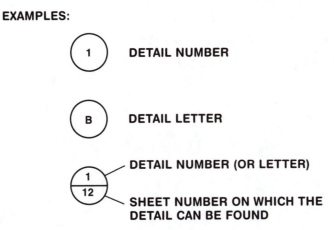

FIGURE 25.3 ■ Identification symbols

DIMENSIONING

Assembly drawings should not be overloaded with dimensions that may be confusing to the print reader. Specific dimensional information should be provided on the detail drawings. Only such dimensions as center distances, overall dimensions, and dimensions that show the relationship of details to the assembly as a whole should be included.

However, there are times when a simple assembly may be dimensioned so that no detail drawings are needed. In such cases, the assembly drawing becomes a working assembly drawing.

ASSIGNMENT D-24: MILLING JACK

1. How many details make up the jack assembly? _____

2. What is the thickness of the ∅ 3.00 section of the base? _____

3. What is the diameter of the boss on the jack base? _____

4. What is the distance from the top of the boss to the top of the base? _____

5. How far is the centerline of the ∅ .625 hole from the centerline of the ∅ .66 hole? _____

6. What is the overall height of the assembled jack when it is in its lowest position? _____

7. What size is the hole in the slide shaft? _____

8. What is the detail number of the slide shaft? _____

9. What is detail ③? _____

10. How many sheets make up the drawing set? _____

11. What material is the jack base made of? _____

12. Of what material is the knurled nut made? _____

13. What is the rough cut stock size of the slide shaft? _____

14. What is the maximum allowable diameter of the jack base? _____

15. What is the size of the tapped hole in the jack base? _____

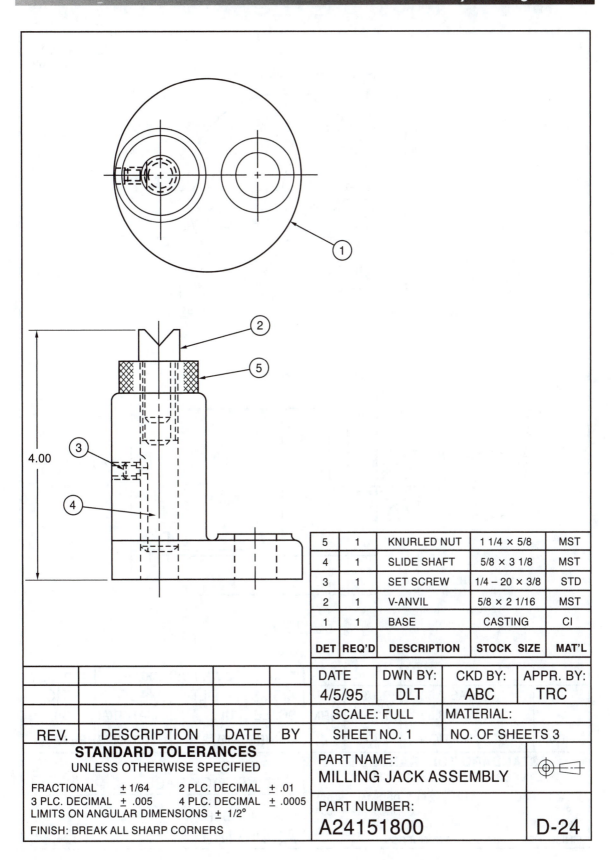

DET	REQ'D	DESCRIPTION	STOCK SIZE	MAT'L
5	1	KNURLED NUT	1 1/4 × 5/8	MST
4	1	SLIDE SHAFT	5/8 × 3 1/8	MST
3	1	SET SCREW	1/4 – 20 × 3/8	STD
2	1	V-ANVIL	5/8 × 2 1/16	MST
1	1	BASE	CASTING	CI

DATE 4/5/95	DWN BY: DLT	CKD BY: ABC	APPR. BY: TRC

SCALE: FULL	MATERIAL:

REV.	DESCRIPTION	DATE	BY	SHEET NO. 1	NO. OF SHEETS 3

STANDARD TOLERANCES
UNLESS OTHERWISE SPECIFIED

FRACTIONAL ± 1/64 2 PLC. DECIMAL ± .01
3 PLC. DECIMAL ± .005 4 PLC. DECIMAL ± .0005
LIMITS ON ANGULAR DIMENSIONS ± 1/2°
FINISH: BREAK ALL SHARP CORNERS

PART NAME:
MILLING JACK ASSEMBLY

PART NUMBER:
A24151800

D-24

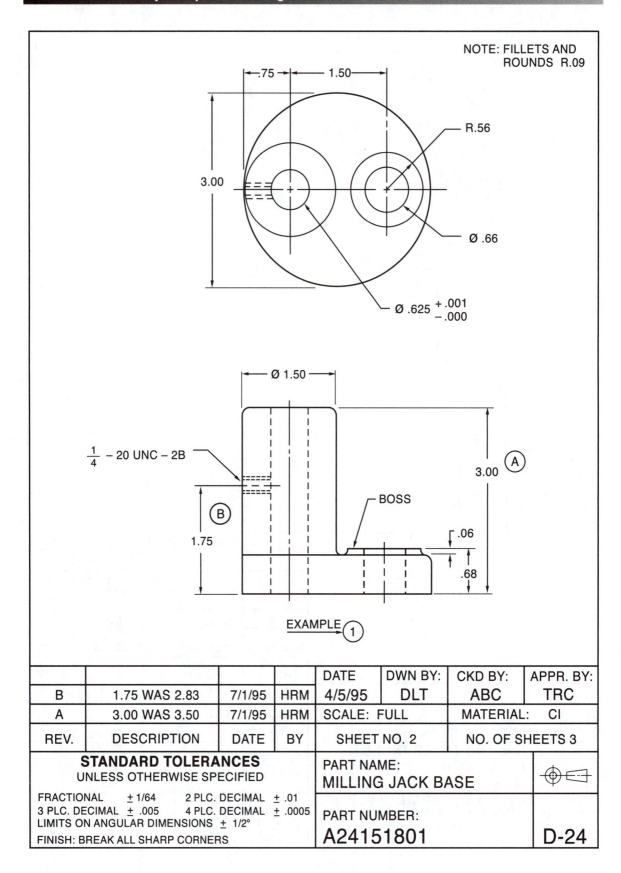

NOTE: FILLETS AND
ROUNDS R.09

.75 — 1.50

3.00

R.56

Ø .66

Ø .625 $^{+.001}_{-.000}$

Ø 1.50

$\frac{1}{4}$ – 20 UNC – 2B

Ⓑ

1.75

3.00 Ⓐ

BOSS

.06

.68

EXAMPLE ①

B	1.75 WAS 2.83	7/1/95	HRM
A	3.00 WAS 3.50	7/1/95	HRM
REV.	DESCRIPTION	DATE	BY

DATE	DWN BY:	CKD BY:	APPR. BY:
4/5/95	DLT	ABC	TRC

SCALE: FULL MATERIAL: CI

SHEET NO. 2 NO. OF SHEETS 3

STANDARD TOLERANCES
UNLESS OTHERWISE SPECIFIED

FRACTIONAL ± 1/64   2 PLC. DECIMAL ± .01
3 PLC. DECIMAL ± .005   4 PLC. DECIMAL ± .0005
LIMITS ON ANGULAR DIMENSIONS ± 1/2°
FINISH: BREAK ALL SHARP CORNERS

PART NAME:
MILLING JACK BASE

PART NUMBER:
A24151801

D-24

V Anvil

1. What size chamfer is required on the anvil? _____
2. How long is the ⌀ .375 diameter? _____
3. How deep is the "V"? _____
4. What are the dimensions for the neck? _____
5. What is the largest diameter for the anvil? _____
6. What is the upper limit dimension for the ⌀ .375 diameter? _____
7. What tolerance is allowed on the 45° angle? _____
8. What is the overall length of the part? _____
9. How many V anvils are required? _____
10. What detail number is the anvil? _____

Slide Shaft

1. What is the depth of the ⌀ .375 hole? _____
2. How long is the 5/8–18 thread? _____
3. What type section is shown at AA? _____
4. What size is the keyseat on the shaft? _____
5. How long is the shaft keyseat? _____
6. What class fit is required on the 5/8–18 threaded section? _____
7. What type of line is shown at Ⓐ ? _____
8. What method of thread representation is shown in the front view? _____
9. The section lining in section AA indicates what type of material? _____
10. What is the lower limit dimension for the ⌀ .625 diameter? _____

Knurled Nut

1. What size knurl is required on the nut? _____
2. How many finished surfaces are required? _____
3. What is the thickness of the nut? _____
4. What size tapped hole is required? _____
5. What is the diameter of the knurled nut? _____

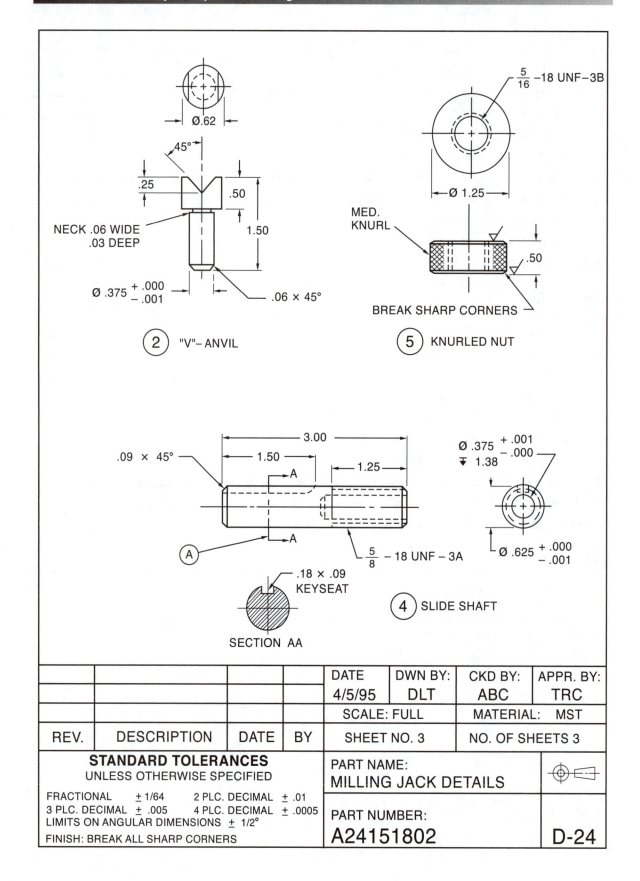

Ø.62

45°

.25   .50

NECK .06 WIDE
.03 DEEP

1.50

Ø .375 $^{+.000}_{-.001}$ .06 × 45°

② "V"– ANVIL

$\frac{5}{16}$ –18 UNF–3B

Ø 1.25

MED.
KNURL

.50

BREAK SHARP CORNERS

⑤ KNURLED NUT

3.00

.09 × 45°   1.50   1.25

A   A

Ø .375 $^{+.001}_{-.000}$
▼ 1.38

A   A

$\frac{5}{8}$ – 18 UNF – 3A

Ø .625 $^{+.000}_{-.001}$

.18 × .09
KEYSEAT

④ SLIDE SHAFT

SECTION AA

				DATE	DWN BY:	CKD BY:	APPR. BY:
				4/5/95	DLT	ABC	TRC
				SCALE: FULL		MATERIAL: MST	
REV.	DESCRIPTION	DATE	BY	SHEET NO. 3		NO. OF SHEETS 3	

STANDARD TOLERANCES
UNLESS OTHERWISE SPECIFIED

FRACTIONAL   ± 1/64   2 PLC. DECIMAL ± .01
3 PLC. DECIMAL ± .005   4 PLC. DECIMAL ± .0005
LIMITS ON ANGULAR DIMENSIONS ± 1/2°
FINISH: BREAK ALL SHARP CORNERS

PART NAME:
MILLING JACK DETAILS

PART NUMBER:
A24151802

D-24

Units 26 and 27 are provided for students who need help in calculating fractional and decimal dimensions on drawings. They include definitions, rules and symbols, and an example of how to apply each rule to a specific problem.

In measuring material, a machinist often uses the major divisions of a steel ruler that are known as inches. The machinist must also know how to use the fractional divisions of an inch, such as 1/2 inch, 1/4 inch, and 1/8 inch. These divisions are also found on the ruler.

The word *digit* refers to the numbers 1, 2, 3, 4, 5, 6, 7, 8, 9, and 0 (zero). In fractional work, as in whole numbers, the following symbols and processes are used:

SYMBOL	WORD	MEANING
+	plus	addition
−	minus	subtraction (difference)
×	multiply	multiplication
÷ x/x $\frac{x}{x}$ x)x̄	divide	division Note that the fraction indicates the process of division.
=	equals	same quantity or value

EXPLANATION OF FRACTIONAL PARTS

If a line is divided into eight parts, one part is $\frac{\text{one part}}{\text{total parts}}$ or 1/8 of the total length of the line. Two parts = 2/8 of the total length; three parts = 3/8; four parts = 4/8; five parts = 5/8; six parts = 6/8; seven parts = 7/8; eight parts = 8/8 or the total length.

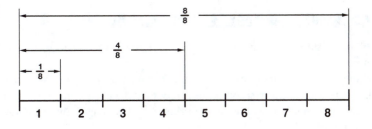

If a line is divided into 16 parts, then one part is 1/16 of the total length of the line. One-half of 1/16 is 1/32 of the total length of the line.

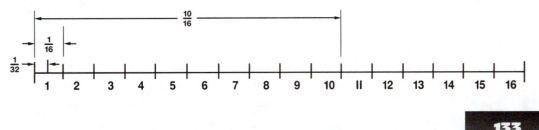

DEFINITIONS OF COMMON FRACTIONS

A *fraction* is a part of a whole quantity, such as 3/4 and 7/32. Thus 3/4 shows that three of the four parts of a whole are considered; 7/32 shows that seven of the thirty-two parts are considered.

The following list contains words and their definitions that apply to fractions:

■ The *denominator* is the figure that shows the number of equal parts into which the whole has been divided. The denominator is written below the line. In 1/2, 2 is the denominator.

■ The *numerator* is the figure that shows how many equal parts of the whole have been taken to make a fraction. The numerator is written above the line. In 1/2, 1 is the numerator.

■ The *terms* of a fraction are the denominator and the numerator.

■ A *common fraction* is a fraction that expresses both terms.

■ A *proper fraction* is a fraction whose numerator is less than its denominator, as 3/8, 5/16, etc.

■ An *improper fraction* is a fraction whose numerator is larger than its denominator, as 3/2, 5/4, 15/8, etc.

■ A *unit fraction* is a fraction whose numerator is one.

■ A *mixed number* is a number composed of a whole number and a fraction, as 3 3/4, 7 1/2, etc.

■ A *complex fraction* is a fraction in which one or both of its terms are fractions or mixed numbers.

Examples: $\dfrac{3/4}{6}$ $\dfrac{15/3}{32}$ $\dfrac{4\ 1/4}{7\ 5/8}$

REDUCTION OF COMMON FRACTIONS AND MIXED NUMBERS

RULE: The value of a fraction is not changed by multiplying the numerator and denominator by the same number. The value of a fraction is not changed by dividing the numerator and denominator of a fraction by the same number.

Examples: $\dfrac{1}{2} = \dfrac{1 \times 4}{2 \times 4}$, or $\dfrac{4}{8}$ Ans. $\dfrac{8}{12} = \dfrac{8 \div 4}{12 \div 4}$, or $\dfrac{2}{3}$ Ans.

When both the numerator and the denominator cannot be divided by the same number, the fraction is in its *lowest terms*. An improper fraction can be reduced to a mixed number by dividing the numerator by the denominator.

Example: 95/8 = 95 ÷ 8, or 11 7/8 Ans.

RULE: A mixed number can be changed to an improper fraction by multiplying the whole number by the denominator and adding the numerator. For example, when changing 2 3/4 to an improper fraction, multiply 2 by 4, add 3, and place the result over 4.

Example: $2\dfrac{3}{4} = \dfrac{(2 \times 4) + 3}{4}$, or $\dfrac{11}{4}$ Ans.

READING THE STEEL RULER

The steel ruler is used to measure and to lay out lengths in inches and fractional parts of inches. The steel rule is also used as a straightedge.

The fractional divisions of an inch are found by dividing the inch into equal parts. The common divisions are halves, quarters, eighths, sixteenths, thirty-seconds, and sixty-fourths. For example, there are eight 1/8-inch divisions in one inch. Inches are designated with a double prime (″). The length, three inches, is written as 3″. The abbreviation for inch, or inches, is *in*.

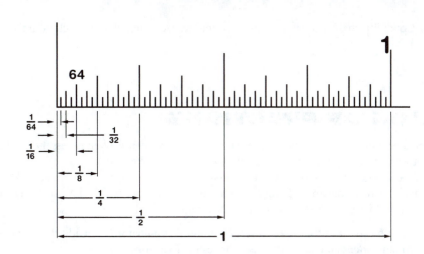

Determine the values of the division before reading a scale. A division is the space between each line. The ruler shown above is graduated in sixty-fourths of an inch. Therefore, each division or space is 1/64″ wide.

8ths Scale. Count three 1/8-inch divisions.

$$A = \frac{3}{8}$$

16ths Scale. Count nine 1/16-inch divisions or count one 1/16-inch division beyond the 1/2-inch (8/16″) division.

$$B = \frac{9}{16}$$

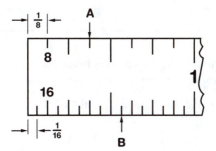

32nds Scale. Count eleven 1/32-inch divisions or subtract one 1/32-inch division from the 3/8-inch (12/32″) division.

$$C = \frac{11}{32}$$

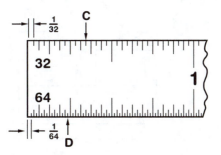

64ths Scale. Count fifteen 1/64-inch divisions or count one 1/64-inch division back from the 1/4-inch (16/64″) division.

$$D = \frac{15}{64}$$

ADDITION OF FRACTIONS

To determine the length of a part from dimensions given on a drawing, it is often necessary to add fractions. *Addition* (+) is the process of finding the sum or total of two or more numbers.

Fractions cannot be added unless they have a *common denominator,* such as 5/8, 7/8, and 12/8. This is a denominator into which all the denominators will divide evenly. It saves work to use the lowest (least) common denominator. For instance, the lowest common denominator of 1/6 and 1/4 is 12.

RULE: To add fractions, reduce them to a lowest common denominator. Then add the numerators, write their sum over the common denominator, and reduce the results to its lowest terms.

Example: Add 3/4, 7/8, and 11/16.

Step 1. Reduce the fractions to the lowest common denominator (16).

$$\frac{3 \times 4}{4 \times 4} = \frac{12}{16} \qquad \frac{7 \times 2}{8 \times 2} = \frac{14}{16} \qquad \frac{11 \times 1}{16 \times 1} = \frac{11}{16}$$

Step 2. Add the fractions by adding the numerators and writing this result over the common denominator.

$$X = \frac{12}{16} + \frac{14}{16} + \frac{11}{16} = \frac{37}{16} = 2\frac{5}{16} \text{ Ans.}$$

Note: When it is possible, reduce the result to a mixed number in its lowest terms.

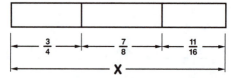

Step 3. *Check:* To check the addition, add the numbers in the reverse direction. To check the accuracy of the fractions 12/16 and 14/16, reduce them to their lowest terms.

RULE: To add mixed numbers, add the whole numbers and the fraction separately, then combine the results.

Example: Add 12 3/4, 14 5/8, and 7 5/32.

Step 1. 12 + 14 + 7 = 33
Step 2. 3/4 + 5/8 + 5/32 = 24/32 + 20/32 + 5/32 = 49/32 = 1 17/32
Step 3. 33 + 1 17/32 = 34 17/32 Ans.

SUBTRACTION OF FRACTIONS

Dimensions that are not on the drawing may be found by subtracting fractional dimensions. *Subtraction* (−) is the process of taking one number from another.

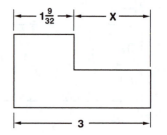

RULE: To subtract fractions, reduce the fractions to the lowest common denominator and subtract the numerators only. Write the difference over the common denominator and reduce the resulting fraction to its lowest terms.

Example: Subtract 1 1/8 from 2 9/16.

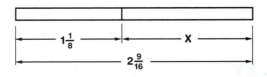

Step 1.   Reduce the fractions to the lowest common denominator.

$$2\frac{9 \times 1}{16 \times 1} = \quad 2\frac{9}{16}$$
$$-1\frac{1 \times 2}{8 \times 2} = -1\frac{2}{16}$$
$$1\frac{7}{16} \text{ Ans.}$$

Step 2.   Subtract the numerators of the fractions: 9/16 − 2/16 = 7/16.
Step 3.   Subtract the whole numbers: 2 − 1 = 1. Thus, 1 + 7/16 = 1 7/16 Ans.

Note: If the numbers are in inches, the answer can be checked by measuring the distance x.

Step 4.   *Check:* The answer can be checked by adding 1 7/16 and 1 1/8. This equals 2 9/16.

RULE: To subtract mixed numbers when the fraction of the mixed number to be subtracted is larger, reduce the fractions to the lowest common denominator, borrow one unit to make up a fraction larger than the one being subtracted, and then subtract in the usual way.

Example: Subtract 4 3/8 from 12 9/64.

Step 1.   Reduce the fractions to the lowest common denominators.
Step 2.   Borrow 64/64 (or 1) from 12 and add this fraction to 9/64.
Step 3.   Subtract the whole numbers and the fractions separately.

$$12\frac{9}{64} = 11\frac{64}{64} + \frac{9}{64} = 11\frac{73}{64}$$
$$-4\frac{3}{8} = -4\frac{24}{64} = \quad -4\frac{24}{64}$$
$$7\frac{49}{64} \text{ Ans.}$$

MULTIPLICATION OF FRACTIONS

Multiplication of fractions is necessary to find the total length of material. *Multiplication* (×) is the process of increasing a number by its own value a certain number of times. It is a short method of adding. The word *of* means multiplication.

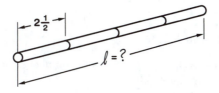

RULE: To find the product of two or more fractions, multiply the numerators and multiply the denominators. Write the product of the numerators over the product of the denominators. Reduce the resulting fraction to its lowest terms.

Example: Multiply 3/4 by 8/9.

Step 1.   Multiply the numerators together and the denominators together. Write the results in fractional form.

$$\frac{3 \times 8}{4 \times 9} = \frac{24}{36}$$

Step 2. Reduce the answer to its lowest terms.

$$\frac{24 \div 12}{36 \div 12} = \frac{2}{3} \text{ Ans.}$$

Step 3. *Check:* Repeat the operations, or use the cancellation method.

CANCELLATION

Cancellation consists of dividing the factors that are common to the numerators and the denominators of the given fractions. *Factors* are parts of a number that, when multiplied together, will give that number. For example, 3 and 2 are factors of 6.

Example: Find the common factors, then multiply 3/4 by 8/9.

Step 1. Divide by 3 to obtain 1 in the numerator and 3 in the denominator.

$$\frac{\overset{1}{\cancel{3}}}{4} \times \frac{8}{\underset{3}{\cancel{9}}} =$$

Step 2. Divide the 4 to obtain 2 in the numerator and 1 in the denominator.

$$\frac{1}{\underset{1}{\cancel{4}}} \times \frac{\overset{2}{\cancel{8}}}{3} =$$

Step 3. Multiply the numerators and the denominators. Write the results in fractional form.

$$\frac{1 \times 2}{1 \times 3} = \frac{2}{3} \text{ Ans.}$$

RULE: Change mixed numbers to improper fractions before multiplying.

Example: $2\frac{1}{2} \times 3\frac{1}{3} = \frac{5}{\underset{1}{\cancel{2}}} \times \frac{\overset{5}{\cancel{10}}}{3} = \frac{25}{3} = 8\frac{1}{3} \text{ Ans.}$

Division of Fractions

Division of fractions is used to find the number of pieces that can be obtained from a length of stock or when it is necessary to divide a line into equal parts.

Division is the process of determining how many times one number is contained in another. Division is indicated by

■ ÷ a division sign, such as 6 ÷ 2

■ a line between the two numbers, such as 6/2 or $\frac{6}{2}$

The *divisor* is the number used to divide. The *dividend* is the number to be divided. The *quotient* is the answer.

Example: In the equation 6/2 = 3: $\frac{6 \text{ (dividend)}}{2 \text{ (divisor)}} = 3$ (quotient)

Note that a fraction expresses the process of division. For example, 4/4 indicates that a unit is divided into four equal parts. The fraction 3/4 indicates that a unit has been divided into four parts and that one of the four parts has been taken away.

When fractions are divided, the problem is to find out *how many* fractional parts are contained in a unit. 3 ÷ 1/2 is another way of expressing "How many halves are there in three?" To find the answer, multiply the whole number by the denominator of the fraction, and then divide that product by the numerator of the fraction. To find the answer, multiply the whole number by the numerator and the denominator of the fraction. This kind of multiplication is called *inversion* because the fraction is turned upside down.

Example: $3 \div 1/2 = \dfrac{3}{1} \times \dfrac{2}{1} = \dfrac{6}{1} = 6$ (There are six halves in three.)

Likewise, when a fraction is divided by another fraction, the dividend (numerator) of one fraction is multiplied by the divisor (denominator) of the other fraction. 7/8 ÷ 3/4 is another way of saying "How many three-quarters are there in seven-eighths?"

Example: $7/8 \div 3/4 = \dfrac{7}{8} \times \dfrac{4}{3} = \dfrac{28}{24} = \dfrac{7}{6}$ or 1 1/6

When a whole number or a fraction is divided by a proper fraction, the answer is always *larger* than the quantity being divided because we use the process of inversion and multiplication.

RULE: To divide fractions, invert the divisor and then multiply the fractions.

Example: Divide 5/7 by 3/4.

Step 1. Write as a division problem.

$$\frac{5}{7} \div \frac{3}{4}$$

Step 2. Replace the division sign with a multiplication sign and invert the divisor.

$$\frac{5}{7} \times \frac{4}{3}$$

Step 3. Multiply as before and reduce the resulting fraction to its lowest terms.

$$\frac{5}{7} \times \frac{4}{3} = \frac{20}{21} \text{ Ans.}$$

Step 4. *Check:* Multiply the divisor (3/4) by the quotient (20/21) to obtain the dividend (5/7), or repeat the operations.

RULE: Mixed numbers should be changed to improper fractions before dividing.

Example: How many pieces of wire 3 1/2 inches long can be obtained from 20 inches of stock?

Step 1. Write as a division problem.
20 ÷ 3 1/2

Step 2. Change the mixed number to an improper fraction.

$$3\frac{1}{2} = \frac{(3 \times 2) + 1}{2} = \frac{7}{2}$$

Step 3. Invert the divisor and multiply: Answer is 5 pieces; the 5/7 is waste.

$$20 \times \frac{2}{7} = \frac{40}{7} = 5\frac{5}{7} \text{ Ans.}$$

USING FRACTIONS IN UNITS OF MEASUREMENT

There are various ways of expressing problems involving the four fundamental operations: addition, subtraction, multiplication, and division. These are indicated with their proper symbols below:

Addition (+)

■ Plus

■ Sum of

■ Added to

■ Increased by

■ Together we have

■ More than

Subtraction (−)

■ Minus
■ Difference between
■ Subtracted from
■ Diminished by
■ Less than

Multiplication (×)

■ Times
■ Product of
■ Multiplied by

Division ($\div$, $\frac{X}{X}$, X/X, X$\overline{)X}$)

■ Divided by
■ Quotient of
■ Fractional part of
■ One number over another

When the four fundamental operations are performed, it is important to know the unit of measurement or the name of the answer. Whenever possible, the unit of measurement should be written after each number as the operations are being performed.

When the unit of measurement is being determined, the following rules will be helpful.

1. When inches or other units of measurements are *added*, the answer will be in the same unit.

   in + in = in

   5 in + 5 in = 10 in

2. When inches or other units of measurements are *subtracted*, the answer will be in the same unit.

   in − in = in

   6 in − 2 in = 4 in

3. When inches or other units of measurements are *multiplied* by a number with no dimension, the answer will be in the same unit.

   in × no = in

   2 in × 3 = 6 in

4. When inches or other units of measurements are *divided* by a number with no dimension, the answer will be in the same unit.

   $\frac{in}{no}$ = in

   $\frac{6 \; in}{3}$ = 2 in

5. When inches or other units of measurements are *divided* by inches or the same unit, the answer will be a number only.

   $\frac{in}{in}$ = no

   $\frac{6 \; in}{3 \; in}$ = 2

Decimal Parts

DEFINITIONS OF DECIMAL FRACTIONS

Dimensions on drawings, particularly tolerances, are frequently expressed in decimal fractions. A decimal fraction is a fraction in which the denominator is 10, 100, 1000, 10000, etc. It is usually not expressed, but signified by a point placed at the left of the numerator.

Examples: $\dfrac{7}{10} = 0.7$ Reads as seven *tenths*.

$\dfrac{7}{100} = 0.07$ Reads as seven *hundredths*.

$\dfrac{7}{1000} = 0.007$ Reads as seven *thousandths*.

$\dfrac{7}{10000} = 0.0007$ Reads as seven *ten-thousandths*.

$\dfrac{7}{100000} = 0.00007$ Reads as seven *hundred-thousandths*.

$\dfrac{250}{100} = 2.50$ or 2.5 Reads as two and *fifty-hundredths* or two and five *tenths*.

Note: The zero before the decimal point emphasizes the fact that the decimal fraction is less than 1.

Value of Decimal Places

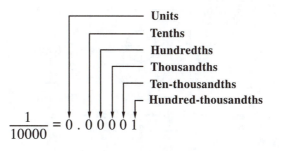

Units
Tenths
Hundredths
Thousandths
Ten-thousandths
Hundred-thousandths

$\dfrac{1}{10000} = 0 . 0 0 0 0 1$

THE DECIMAL RULER

There are rulers that are divided according to the decimal scale. The drawing on page 142 shows a ruler divided into 50ths and 100ths of an inch. Such scales reduce the possibility of error that occurs when common fractions are changed to decimals.

Examples: $\dfrac{1''}{10} = 0.1'' =$ one-tenth of an inch

$\dfrac{1''}{50} = 0.02'' =$ one-fiftieth of an inch

$\dfrac{1''}{100} = 0.01'' =$ one-hundredth of an inch

$\dfrac{1''}{1000} = 0.001'' =$ one-thousandth of an inch

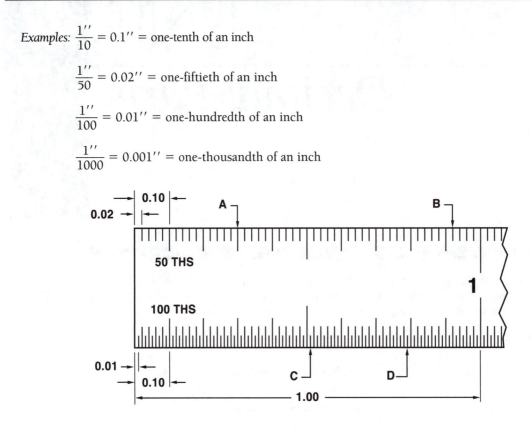

To read the distances from the end of the ruler (above) to points A, B, C, and D, proceed as follows:

A = 0.3″   Count three 1/10-inch divisions.

B = 0.9″ + 0.02″ = 0.92″   Count nine 1/10-inch divisions and add the value of the smallest division.

C = 0.5″ + 0.01″ = 0.51″   Count five 1/10-inch divisions and add the value of the smallest division.

D = 0.7″ + 0.09″ = 0.79″   Count seven 1/10-inch divisions and add the value of nine of the smallest divisions, or count one small division back from 0.8″.

ADDITION AND SUBTRACTION OF DECIMALS

When dimensions are expressed in decimal form it is often necessary to add or subtract them to find the total length or the distance between certain points on a job.

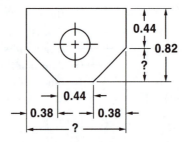

Note: The instructor may allow the use of a calculator to determine the answers to the following math problems.

RULE: To add or subtract decimals, place the numbers so that the decimal points are under each other. Then add or subtract as with whole numbers.

Example: Add 3.25, 864.0725, 647, and 0.875.

Step 1.   Arrange the given numbers so that the decimal points are under each other.

$$
\begin{array}{r}
3.25 \\
864.0725 \\
647. \\
+ \quad 0.875 \\
\hline
\end{array}
$$

Step 2. To avoid errors, add zeros to the numbers with the fewer places so that the numbers have an equal number of places after the decimal point.

$$
\begin{array}{r}
3.2500 \\
864.0725 \\
647.0000 \\
+ \quad 0.8750 \\
\hline
\end{array}
$$

Step 3. Proceed as in addition of whole numbers. Place the decimal point in the sum under the other decimal points.

$$
\begin{array}{r}
3.2500 \\
864.0725 \\
647.0000 \\
+ \quad 0.8750 \\
\hline
\end{array}
$$

(Sum)   1515.1975 Ans.

Step 4. *Check:* Add the numbers in reverse order.

Example: From 42.63 take 18.275.

Step 1. Write the numbers so that the decimal points are under each other and the subtrahend is under the minuend.

(Minuend)   42.63
(Subtrahend)   − 18.275

Step 2. If one of the numbers has fewer figures after the decimal point than the other, add zeros to it.

(Minuend)   42.630
(Subtrahend)   − 18.275

Step 3. Subtract as with whole numbers. Place the decimal point in the remainder under the other decimal points.

$$
\begin{array}{r}
42.630 \\
- 18.275 \\
\hline
\end{array}
$$

(Remainder)   24.355 Ans.

Step 4. *Check:* Minuend equals sum of subtrahend and remainder.

MULTIPLICATION OF DECIMALS

The multiplication of decimals is used for finding the total length of material needed for a job.

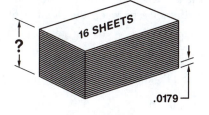

RULE: To multiply decimals, multiply as in whole numbers. Point off as many decimal places in the product as there are decimal places in both the multiplicand and the multiplier.

Example: Multiply 43.286 by 6.04.

Step 1.   Write the number with the most digits first.

(Multiplicand) 43.286
(Multiplier)   × 6.04

Step 2.   Multiply as with whole numbers.

43.286
× 6.04
173144
2597160
(Product)   26144744

Step 3.   Beginning at the right, point off in the product the sum of the decimal places in the multiplicand and the multiplier.

(3 places)   43.286
(2 places)   × 6.04
173144
2597160
(5 places) 261.44744 Ans.

Step 4.   *Check:* Either interchange the multiplier and multiplicand and rework the problem, or repeat the operations.

Example: Multiply 0.1875 by 0.02.

Multiply. Prefix two zeros to make the necessary six places. Drop the zero to the left of the decimal point.

(4 places)   0.1875
(2 places)   × 0.02
0.003750 Ans.

Note: When multiplying numbers such as 25.020 by 4.690, drop the zeros at the extreme right.

25.020 =   25.02
× 4.690 = × 4.69

DIVISION OF DECIMALS

The division of decimals is used to find the number of smaller pieces that can be obtained from a whole piece.

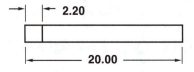

RULE: To divide decimals, divide as in whole numbers. Point off in the quotient (answer) the number of decimal places equal to the difference between the number of decimal places in the dividend and those in the divisor.

Example: Divide 0.78 by 0.964 to 4 decimal places.

Step 1.   Place the dividend inside and the divisor outside the symbol as shown.

(Divisor) 0.964 $\overline{)0.78}$ (Dividend)

Step 2.   Make a whole number out of the divisor by moving the decimal point and use a caret (∧) to indicate the new decimal place.

$$0.964_\wedge \overline{)0.78_\wedge}$$

Step 3. Move the decimal point in the dividend the same number of places to the right, indicating the position with a caret. If necessary, add zeros to the dividend.

$$0.964_\wedge \overline{)0.780_\wedge}$$

Step 4. Divide as in whole numbers, adding as many zeros after the caret in the dividend as are necessary to give the required number of places in the quotient. Place the decimal point in the quotient directly above the caret in the dividend.

$$
\begin{array}{r}
0\ .8091 \text{ Ans.} \\
0.964_\wedge \overline{)0.780_\wedge 0000} \\
\underline{771\ 2} \\
8\ 800 \\
\underline{8\ 676} \\
1240 \\
\underline{964} \\
.0000276
\end{array}
$$

Step 5. *Check:* Multiply the quotient by the divisor and add the remainder, or repeat the operations.

Note: If there are extra zeros in the divisor, drop these before proceeding with the problem. If 0.20900 is the divisor, use .209.

ROUNDING OFF DECIMALS

To save time in working a problem, no more decimal places should be held than are indicated by the accuracy of the measurements. In order to hold the decimal places to a required number, the numbers must be rounded off. *Rounding off* means expressing a number with fewer digits. For example, 0.15934 inch rounded off to 3 places is 0.159 inch.

RULE: Decide how many decimal places are needed in the answer. If the number following the last decimal place required is 5 or larger, add 1. If this number is less than 5, do not add 1. Drop all numbers following the last decimal place required.

Example: Round off 0.672853 to three decimal places.

Step 1. Three decimal places are required.
Step 2. The number following the last decimal place required is 8; therefore, add 1 to the 2, and .672 becomes .673.
Step 3. Drop 0.000853, and the final result is 0.673 Ans.

NUMBER OF PLACES IN COMPUTATIONS

Hold one or more decimal places during the computations than the number of decimal places required in the answer. For instance, if the measurements are correct to three places, hold four places throughout the computations and three places in the answer. This is a good general rule to follow. The exceptions to this rule will be given later.

Example: Multiply 12.125 by 2.510 and then by 6.875. Three decimal places are required in the answer.

$$
\begin{array}{r}
12.125 \\
\times\ 2.51 \\
\hline
12125 \\
60625 \\
\underline{24250} \\
30.43375 = 30.4338
\end{array}
\qquad
\begin{array}{r}
30.4338 \\
\times\ 6.875 \\
\hline
1521690 \\
2130366 \\
2434704 \\
\underline{1826028\ \ \ \ } \\
209.2323750 = 209.232 \text{ Ans.}
\end{array}
$$

CONVERSION OF COMMON FRACTIONS AND DECIMAL FRACTIONS

It is often necessary to change dimensions expressed in the form of common fractions to dimensions expressed in decimal form, or vice versa. When the denominator of a fraction can be evenly divided *into* 100, or evenly divided *by* 100, the following method may be used for conversion.

Example: Change a fraction into a decimal.

Step 1.   Denominators less than 100: Change 3/25 to hundredths.
    a.   Divide the denominator 25 into 100.

$$\frac{100}{25} = 4$$

    b.   Multiply the numerator and the denominator by the result (4).

$$\frac{3 \times 4}{25 \times 4} = \frac{12}{100} = 0.12 \text{ Ans.}$$

Step 2.   Denominators greater than 100: Change 8/200 to hundredths. Divide the denominator and the numerator by the number needed to give 100 in the denominator.

$$\frac{8 \div 2}{200 \div 2} = \frac{4}{100} = 0.04 \text{ Ans.}$$

When the above method cannot be used, multiplication and division according to the procedures for decimal values is done. These methods can be used to change other common fractions and decimals such as those involved in dollars and weights.

CHANGING A COMMON FRACTION TO A DECIMAL FRACTION

RULE: Divide the numerator of the common fraction by the denominator. Add zeros after the decimal point, if necessary.

Example: Change 15/16 to a decimal fraction.

Step 1.   Write the common fraction as a division problem.
Step 2.   Since 16 does not go into 15, place a decimal point after the 5, add zeros, and divide. In this case, the division was continued until there was no remainder.

```
      0.9375  Ans.
16)15.0000
   14 4
      60
      48
     120
     112
       80
       80
```

CHANGING A DECIMAL FRACTION TO A COMMON FRACTION

RULE: Multiply the given decimal by the fractional divisions required. Form a common fraction with this result (rounded off) as a numerator and the divisions required as a denominator.

Example: Change 0.385 to a common fraction (closest 64ths).

Step 1. Multiply the given number by the closest divisions needed; in this case by 64. The result is 24.64.

$$\begin{array}{r} 0.385 \\ \times\ 64 \\ \hline 1540 \\ 2310 \\ \hline 24.640 \end{array}$$

Step 2. Round off 24.64 to a whole number.

24.64 or 25

Step 3. Put this result (25) over 64.

$\dfrac{25}{64}$ Ans.

Check: 25/64 = 0.390625
24/64 = 0.375

APPENDIX

STANDARD ABBREVIATIONS*

acme screw thread	ACME	forged steel	FST
allowance	ALLOW	gauge	GA
alloy steel	ALY STL	galvanized	GALV
aluminum	AL	grind	GRD
anneal	ANL	harden	HDN
approved	APVD	heat treatment	HT
as required	AR	high-carbon steel	HCS
bevel	BEV	hot-rolled steel	HRS
bracket	BRKT	inch	IN
brass	BRS	inside diameter	ID
brazing	BRZG	keyway	KWY
break	BRK	laminate	LAM
Brinell hardness number	BHN	left hand	LH
bronze	BRZ	length	LG
burnish	BNSH	lubricate	LUB
bushing	BSHG	machine steel	MST
case harden	CH	magnesium	MAG
cast iron	CI	malleable	MAL
center	CTR	malleable iron	MI
chamfer	CHAM	material	MATL
chrome vanadium	CR VAN	maximum	MAX
clearance	CL	millimeter	MM
cold-drawn steel	CDS	minimum	MIN
cold-rolled steel	CRS	national coarse thread	NC
copper	COP	national extra-fine thread	NEF
counterbore	CBORE	national fine thread	NF
countersink	CSK	nickel steel	NS
cylinder	CYL	nominal	NOM
degree	DEG	not to scale	NTS
diagonal	DIAG	number	NO
diameter	DIA	outside diameter	OD
dimension	DIM	plastic	PLSTC
drawing	DWG	quantity	QTY
drill	DR	radius	R or RAD
fabricate	FAB	ream	RM
fillet	FIL	reference	REF

*American National Standards Institute

finish	FIN	required	REQD
finish all over	FAO	right hand	RH
flange	FLG	round	RD
Society of Automotive		thread	THD
Engineers	SAE	tolerance	TOL
spotface	SF	tool steel	TS
symbol	SYM	tungsten	TU
tangent	TAN	typical	TYP
tapping	TAP	unified national course	UNC
tensile strength	TS	unified national extra-fine	UNEF
thick	THK	unified national fine	UNF

INDEX